SpringerBriefs in Architectural Design and Technology

Series editor

Thomas Schröpfer, Architecture and Sustainable Design (ASD), Singapore University of Technology and Design, Singapore, Singapore

Understanding the complex relationship between design and technology is increasingly critical to the field of Architecture. The *SpringerBriefs in Architectural Design and Technology* series provides accessible and comprehensive guides for all aspects of current architectural design relating to advances in technology including material science, material technology, structure and form, environmental strategies, building performance and energy, computer simulation and modeling, digital fabrication, and advanced building processes. The series features leading international experts from academia and practice who provide in-depth knowledge on all aspects of integrating architectural design with technical and environmental building solutions towards the challenges of a better world. Provocative and inspirational, each volume in the Series aims to stimulate theoretical and creative advances and question the outcome of technical innovations as well as the far-reaching social, cultural, and environmental challenges that present themselves to architectural design today. Each brief asks why things are as they are, traces the latest trends and provides penetrating, insightful and in-depth views of current topics of architectural design. *Springer-Briefs in Architectural Design and Technology* provides must-have, cutting-edge content that becomes an essential reference for academics, practitioners, and students of Architecture worldwide.

More information about this series at http://www.springer.com/series/13482

Andrea Quartara · Djordje Stanojevic

Computational and Manufacturing Strategies

Experimental Expressions of Wood Capabilities

 Springer

Andrea Quartara
Genoa, Italy

Djordje Stanojevic
Guadalupe, Nuevo León
Mexico

ISSN 2199-580X   ISSN 2199-5818 (electronic)
SpringerBriefs in Architectural Design and Technology
ISBN 978-981-10-8829-2   ISBN 978-981-10-8830-8   (eBook)
https://doi.org/10.1007/978-981-10-8830-8

Library of Congress Control Number: 2018951903

This Springer imprint is published by the registered company Springer Nature Singapore Pte Ltd.
The registered company address is: 152 Beach Road, #21-01/04 Gateway East, Singapore 189721,
Singapore

Preface

This book is directed towards architecture students and practitioners who are interested in exploring both design-to-fabrication opportunities and challenges. It intends to establish a logical design workflow that is valuable on an educational level and that can stimulate designers' creativity. Computational and Manufacturing Strategies is based on the Ph.D. thesis titled "Post-Digital Reflections" by A. Quartara, submitted to the Department of Architecture and Design (DAD) at the University of Genova (UNIGE) in 2017 as well as the research by D. Stanojevic on material enhancement and its computational form conducted over the last 2 years. The book presents work related to digital design, CNC fabrication and wood as a construction material.

Computational and Manufacturing Strategies first introduces theoretical foundations and subsequently focuses on the possibilities that emerge from material-aware design processes. The first part of the book presents the selected important contributions from the many that have played an important role in this field of research. These are presented in the form of a story that the authors believe will be valuable particularly for students of architecture. The first chapter outlines a speculative background and experiential projects are then discussed from the vantage point of the customization of digital production that is enabled by new machines and the logic of the construction material itself. Technologically innovative procedures and manufacturing strategies are combined in order to explore and extend architectural enquiries. Nevertheless, the primary interests concern the exploration of different fabrication methods to deal with tolerances of raw and engineered material. On the other hand, they address studies on material enhancements and strip topologies applied in timber construction. The pavilions presented in the book demonstrate that wood is one of the most suitable materials to allow a full integration of the digital design-to-construction workflow for a seamless framework on an educational level.

Genoa, Italy
Guadalupe, Mexico

Andrea Quartara
Djordje Stanojevic

Acknowledgements

The presented case studies were developed within international academic groups from the Institute for Advanced Architecture of Catalonia (IAAC), the Architectural Association's Hooke Park Campus and the Centro de Estudios Superiores de Diseño de Monterrey (CEDIM). The full-size wooden structures are considered an extensible experiential learning framework, which acts as an interoperable procedure, bridging virtual and real.

For this, the authors would like to thank all the talented figures involved in the pavilions built in Barcelona: A. Markopoulou, for her inspiring and critical thoughts, A. Dubor for his clever tutoring and vast knowledge-sharing, and S. Brandi and M. Kuptsova for their work as coordinators (all: IAAC, Barcelona). Equally, the authors would like to express their gratitude to M. Seymour (Fab Lab Barcelona Manager) and R. Valbuena (Fab Lab Barcelona) for their suggestions during fabrication and A. Pistofidou (Fab Textiles Project Leader) for her insightful advice. The authors acknowledge the motivating project reviews during the Open Thesis Fabrication research by E. Ruiz-Geli. Projects at IAAC would not have been developed without the collaboration of Serradora Boix, Setmana de la fusta de Catalunya, Gremi de Fusters, Tallfusta, Incafust, Mecakim, Decustik, Windmill and Merefsa.

A special thank you goes to E. Vercruysse, the director of the Robotic Fabrications workshop, for his passion in sharing his in-depth experience of both traditional and digital craft; to G. Edwards (collaborator), P. Devadass (robotic developer), and Z. Mollica (collaborator) for their strauch effort in assisting the workshop; to Martin Self (Hooke Park director) for his striking dialogues, and to Charlie Corry-Wright for his essential help during assembly procedures (all: Hooke Park Campus, Architectural Association).

In addition, the authors would like to thank G. Kazlachev for collaborating in the theoretical conceptualization, computational design development and tutoring for the realization of the Woven Wood and Synthesis of Strip Pattern structures. The two prototypes have been realized during workshops promoted and organized by Noumena, hosted by Sbodio32, Nodo and IAAC. We are really thankful to have been supported by the team from Karamba3D, who provided software licenses for the workshops. We wish to acknowledge R. Aguirre for his helpful advice with computational design approaches on these two projects. The authors would like to express their gratitude towards D. Durán Sánchez, Academic Director of the Architectural Department from the Centro de Estudios Superiores de Diseño de Monterrey (CEDIM) for supporting the research of the presented work over more than a year including The Laminate Pavilion and ongoing projects. Thanks to A. L. Scherer for proofreading the content.

Authors are grateful to all the tutors, researchers and students for the effort in working side by side (in alphabetical order): S. Akcicek, J. Alcover, S. Almorelli, E. Azadi, M. Bannwart, C. Bertossi, J. Blathwayt, G. Bruni Zani, P. Bussold, A. Carpenter, J. E. Castillo Neyra, Z. E. Cepeda Avila, B. Chavez, J. Curry, S. Cutajar, P. De Groeve, I. Di Stefano, L. E. Doster Arizmendi, I. Durán Vigil, M. Esquivel Garza, A. Figliola, D. Fiore, G. Galli, D. Garcia Pando, A. Giacomelli, A. Giglio, Y. Haddad, Ka. Kaewprasert, Ko. Kaewprasert, A. Kastorinis, M. Kumar, Y. Li, M. Magdy, S. Martinez Jiménez, F. M. Massetti, S. Meloni, C. Mendoza, A. Meza Murcia, M. M. Najafi, M. Orozco, G. Pernsilici, M. Pilon, S. Rademeyer, A. Revelles Elizondo, N. M. Rigal Delgado, E. Rodionov, E. Saccaperni, J. Sainz de Aja Curbelo, D. Saldivar Elizondo, I. F. Sandoval Reyes, F. S. Shakir, N. Shalaby, M. Sharp, Z. Sun, D. Tamez, C. Thompson, E. Triantafyllidou, S. Varani, E. Vasileska, M. A. Villarreal Muñoz, T. Wang, J. Won Jun, V. Winterle, A. Xristov, Y. Yang and E. Zanetti.

Last, but not least, our greatest loving gratitude goes to our relatives, for their continuous love and support.

Contents

Chapter 1
Know and Build Your Own Tool

1.1 Features

Technological advancement can be certainly recognized as the mainstay of every evolutionary stage of mankind, however primitive or sophisticated it may be. Material means and cognitive tools in here, therefore, any culture and consistently are producing improvements on how humankind achieves results. This is evident if we define the generic tool as the most fundamental level of idea's emergence—namely invention—that has conclusively conditioned the mankind evolution. And thus, a tool, or a toolkit, is not only something that we hold in our hands and use to do a particular job, as every dictionary definition states. It is not simply a piece of equipment for getting the user to perform a task, but it is foremost the set of skills and knowledge useful for doing that job. It is something that has to be seized, known and perhaps built from scratch, becoming a seed that can produce unpredictable uniqueness when it is adopted and adapted while facing a task. Historically, architects tend to be late in embracing technological changes: common features borrowed from everyday life evolution become central and distinguishing ones when architect are acquainted with and start to alter them according to new tasks they are exploring. In the last 40 years, the advent of digital turns comes to be distinctive features for architects, and this is why computer logics have been pivotal for architectural design. According to M. Carpo's investigation—emerging from his book production between the 2013 and the 2017—architects have witnessed two main digital turns in the last three decades; they are a consequence of technological advancement adopted and adapted from time to time. On the one hand, the early generation of digital architects has initiated the experimentations and productions made of splines, NURBS and blobs defining an iconic identity for the first digital age. On the other hand, the occurring post-digital[1]

---

[1] During the reading of the book the definition that the author wants to attribute to post-digital expression will emerge. For now, the reader is aware of an antinomy between digital and material that must be discussed in today's architectural debate.

© The Author(s) 2019   1
A. Quartara and D. Stanojevic, *Computational and Manufacturing Strategies*,
SpringerBriefs in Architectural Design and Technology,
https://doi.org/10.1007/978-981-10-8830-8_1

opportunity drives forward the groundbreaking IT logics: by replacing deduction from mathematical formulas with form-finding processes and directly embedding the making process within the digital design flow, the current digital academic generation is leading experimental researches. Such emergence cannot be dissociated from tool and material features involved in the design process, enabling architects to practice the file-to-factory seamless venture.

1.1.1 Tool Features

Within the context of this narrative, the tool is intended as the medium through which architects devise space. Only afterwards, they address the ideation process to its materialization well supported by technological means (both software or hardware). And then, a toolbox for architects is here equipped with a set of influential theoretical knowledge, or guiding reasons, along with a series of fabrication machines. The complementary role between the soft outcome—namely the design insight—and its hard materialization characterizing space is seen as the symbolic transfer of the aforementioned relation between a system of knowledge and the tool. This intertwinement is clearly expressed by M. Weiser in his work *The Computer for the 21st Century* (Weiser 1991):

> [...] The most profound technologies are those that disappear. They weave themselves into the fabric of everyday life until they are indistinguishable from it.

The objective of the following sections is to seek the origin of these disappearing tools that are implemented within the described design-to-construction architectural researches. Therefore, the narrative depicts a possible genesis of the close interrelation between computational thinking and fabrication methods within a non-virtual framework. The assumption that digital comprises both a chance and a challenge to our discipline makes valuable the attempt to establish a wider overlay of theoretical principles adopted from the most disparate disciplines in order to validate the practical applications presented in the *Material: Digital in Action* chapter. This confluence of fields ranging from mathematics, Information Technology (IT), biology and philosophy, material science and traditional manual craftsmanship constitutes the profound architectural potential of digital implementations. In order to develop central disciplinary concerns, architects have to know and set-up their own computational and productive tools. New technology is confirmed as an "enabling apparatus" (Kolarevic 2003), as it has been for the medieval master builders; pervasive digital technologies end up awakening the demand of architects to be reinvolved in the act of building.

The narrative approach delves this pair deeper by a critical[2] reasoning. On the one hand, the symptomatic title of S. Giedion's book—*Mechanization Takes Com-*

---

[2]The word critical is here used by its true etymological meaning; from Greek *krinô*, meaning distinguish, choose and/or evaluate.

mand, dated back to 1948[3]—can be paraphrased to characterize the investigation. Machines take command: the belief that stands at the origin of the inquiry is that technological innovations—including personal computer that makes complex computational software work and computer numerically controlled (CNC) machines that manipulate material—have guided and continue to guide architects' works. The narrative discerns how far digital turn corresponds with and to what extent it reorganizes architectural conceptions. Digital media have aroused a paradigmatic change of perspective in architecture: digital machines, as computer are, have produced a real Copernican revolution empowering the so-called bottom-up logic. In Le Corbusier's masterwork *Vers une Architecture*, he argued that the world's transformation by machines would lead either to "an amelioration, of historical importance" [or to a] "catastrophe".

On the other hand, while taking command, technological innovations also deploy the question of their limits. Ever-increasing power and quality of technological tools unavoidably may present weaknesses. Fallacy is bound to arise at any moment but even so, it represents a trigger or a starting point. As G. Scott observes in his book, *The Architecture of Humanism A Study in the History of Taste*, published in 1965, architecture and humans had met several fallacies over the centuries. In the following case studies, several shortcomings have necessarily occurred, mostly relating to form-finding and physical manufacturing. They have been explored by researchers and have been interpreted as stimuli for the customization of productive tools with ameliorative features.

These kinds of custom tools are involved as design parameters within the computational chain and the following sections come to envision computation as a practice of designing as well as a practice of materialization. Their aim is to depict an investigative path, which supports case studies. Digital fabrication[4] is promoted as the real accomplishment that leads the forms-in-potency of the aforementioned Copernican revolution, to their actuality,[5] bringing the broad spread of logic and symbolic computation in architecture to the size of digital manipulation of materials.

In this category of tools feature, we include the personal computer as that generic tool, made available to all, which is able to adapt thanks to several sets of software. 60 years ago, the first software for computer-aided design and manufacturing

---

[3]Other interesting publications have already adapted Giedion's expression. L. Manovich has written *Software Takes Command: Extending the Language of New Media* in 2013 and G. Barazzetta has edited a book titled Digital Takes Command in 2015, including several contributions displaying technologies that permit the passage from ideas and figurations to the material reality. K. Moe wrote the chapter *Automation Takes Command*, retrieved in the book *Fabricating Architecture* by R. Corser (2010) looking at the history of technical developments in fabrication technologies.

[4]In particular, robotic manufacturing as it is more interesting for architects as it makes the 1:1 process scalability efficient.

[5]The following paragraph—*Forms-In-Potency versus Forms-In-Actuality*—clarifies this statement, bringing authoritative sources to explain the terms used.

(CAD/CAM)[6] allowed architects to implement geometric constraints in the act of drawing. The first use practices clearly demonstrated how computer graphics could be used for both artistic and technical purposes in addition to showing the novel method of human–computer interaction. And so before long, the ever-increasing availability of computing power has allowed it to free itself from applications on the Cartesian plane, and advanced IT practices have become exploration topics for architects.[7] Data processing and computational procedures help to transfer and simplify design problems using mathematical notation: Bezier's splines and NURBS are recognized as the two main actors of the first digital revolution occurred during the nineteenth century. The complexity of shapes has been often condensed within an elegant and streamlined math notation, that is, rational problem-solving. Computer can repeatedly run calculus—generation, optimization and simulation—also implementing structural or environmental rules in a new scientific way, that is form finding. It is a digital trial-and-error method that supports designers in the choice of a single solution out of the numerous alternatives that computer can generate. It is a making and re-making design options without even building one single part of it, but achieving the best solution on the screen. We can draw a parallel with the traditional craftsman practice: a tradesman does not calculate the best shape of the piece but is searching for the best form for the chair by trial and error.[8] The toolkit's advancements provide architects with always-updated instruments to produce geometries responding to project's requests. Nevertheless, also other important features contribute to post-digital architecture.

1.1.2 Material Features

In recent decades, the significant advances in CAD have emphasized the role that geometry has always played in architectural investigations. Often, these profound changes have produced only instrumental geometries, missing crucial aspects such as feasibility and constructability.[9] Make is central to the act of design and make certainly requires material.

The making for architect, as it happens for carpenters or craftsmen, deals with material and machines, and then with their features, both constraints and virtues. Besides, digital-based mechanistic processes, material claims the identity of being a tool itself for designers. And this book chooses wood as the material to show how

---

[6]Sketchpad, the revolutionary computer program written by I. Sutherland in 1963 during his PhD thesis at MIT, can be considered the ancestor of modern computer-aided design (CAD) programs as well as a major breakthrough in pointed forward to object oriented programming.

[7]Such as nonlinearity, double curvature surface's manipulation, scripting, optimization, up to the recent information modelling and parametric implementations.

[8]Experience and intuition helps the craftsman in achieving the final product.

[9]Constructability and buildability are uncommon words expressing the degree to which the integration of experience and knowledge in a construction process facilitates achievement of an optimum balance between project goals and resource constraints.

in different experimentations only combinatory design can emphasize noteworthy material configurations.

Besides, the unique but intangible virtual ability of digital design, many pioneering practitioners have steered the passage from the early days of computing to the digital architecture as a matter of production innovation. The transfer of software knowledge to the material field allows in really implementing the computational design. In fact, according to its Latin etymology *cum* and *putàre*, computation literally means put together: the calculus power undoubtedly extends the limits of geometric complexity, but it really leads to digital architecture when combined with material investigation. Digital design and digital fabrication increasingly converge, sharing programming protocols and virtual environment. When investigating material features as parameters to be expanded and as a tool to be exploited during each design phase, new opportunities for designers and new horizons for the building industry are arising.

A material practice is crucial at any stage of design; manual experiences and the direct contact with matter encourage architects to call into question their drawings. Sometimes it happens sketches and illustrations run up against simplifying design problems, representing bidimensional lines onto blank sheets of paper while considering the physical aspect of each project issue is the only way to face the design challenge. The attitude of taking in account material features as designing elements—when combined with the computational power during experimental projects as the ones later described—makes a decisive enhancement. An important feature that matter brings with it is that of inspiring students, researchers and practitioners to go beyond the conventional rules of representation that often relegate the materiality of a project to different hatches or notations. The lack of material attention is demonstrated by professional practice where

> the split appears again in the tender package, where drawings describe form but language is used for the materials of building in notes on working drawings and in the specification.
>
> (Thomas 2007)

The previously described technological changes occurred in practice provide the conceptual framework with computer numerically controlled (CNC) and rapid prototyping (RP) processes which make a direct link between ideation and production, merging the form/matter split that has appeared within the discipline over time possible. The diffusion of interest about matter in architecture lays its foundations in different areas of endeavour, from technological advances, environmental concerns and also theoretical debates. And instead of being focused on shaping aspects, with this storytelling, we look at materials towards likewise crucial questions of practice, therefore, redirecting the formal ones.

1.2   Forms-in-Potency Versus Forms-in-Actuality

> The manifest form - that which appears - is the result of a computational interaction between
> internal rules and external pressures. [...] internal rules comprise, in their activity, an embedded form.
>
> (Kwinter 2003)

S. Kwinter with his words seems to reinstate the ancient Aristotle's view expressed in *Metaphysics* books. It is clear the direct link between the substance of the thing, defined as *synolon* (from Latin, meaning the whole thing or the totality) by the philosopher, and the manifest form here outlined. According to the Greek thinker, the set of form and matter, which cannot be separated, as logical components, originates the mutation: when matter has passed to a certain form it is said that it has passed to the act. Hence, matter is power, form is act. Mutation is the actual implementation of what is potential. Aristotle explains the becoming not as a passage from non-being to being, but from being in a certain way—what we define in potency—to being in another way—in activity in our case. In the perspective that material and idea are part of a continuum of potentiality and actualization, the New York-based architectural theorist relates the manifest and material form with the computational ability to pair internal rules with external influences. Here, we intend internal rules as the design process, which establishes a logic order between the involved features,[10] and the external rules the required material procedures in order to produce tangible forms.

This specially-made reinstatement turns the tide here; the link between abstraction and workability, which entails a close connection between processing thoughts and tangible action is one of the most important prerequisites for digital architecture characterizing its introduction in all sectors as the case studies take on.

Architects cannot disregard forms' representation, whether they are "embedded" or "manifest" (Kwinter 2003); sketches, plans, sections, elevations and models are unavoidable media through which architects represent an immaterial substance: the space. They describe ordered conceptions throughout graphical information intended to ask other disciplines to solve problems, most of the times associated with the construction. We can consider the everlasting dialogue between architecture, engineering and construction, occurring by means of drawings, as another transfer of the previous potential form—that is drawn down by architects—and the actual one—its construction. Over the centuries, representations have appeared as engraving, hand-drawing, xylography, lithography and ink print up to be paperless information. And to some extent, they have always been digital representations.[11] But that entire aside, we are interested in describing how the translation from potential forms and actual constructions occurs. Indeed, we can make our own the C. Alexander's attitude in his essay published in 1977; he introduces his renowned expression of a pattern language that we can ascribe to an architectural drawing whereby designers define a problem and

---

[10]The reader is suggested to consider this term in the dual meaning defined in the previous chapter.

[11]In fact if we look at the Latin origin of word, we see that digital comes from *digĭtus*, meaning finger. Architects' hands are the ultimate media used to express potential design outcome.

then offer a solution. The book provides the reader with 253 models, which gather a paradigmatic resolved system of knowledge. But we can also take inspiration from his books in order to outline some literary bases of this firm link.

If we go looking for the etymological parents of architectural shape, we definitely notice a pattern as one word that claims with its roots—from Latin *pater*, meaning father—an authoritative role. In the same manner, we can undoubtedly connect the material to its Latin origin *mater,* meaning mother and thus state that a material pattern accommodates this figurative generation.

> Patterns' newfound integrity in architecture can be explained to some degree [...] as a series or sequence of repeated elements. This has made architects realize that patterns [...] are intrinsic

To the computational design path, as H. Castle expresses in introducing the issue number 6 of *Architectural Design* in 2009 titled *Patterns of Architecture.* Architectural form, strictly conceived as conceptual or cerebral construct, interprets material as inert receptacles for a superimposed order. But in the light of the design's meaning here encouraged, materials "are active participants in the genesis of form" as suggests the philosopher M. DeLanda in his article *Philosophies of Design: The Case of Modelling Software* included in VERB in 2002.

Hence, the duo pattern-material provides architectural form with different but essential features: on one side patterns are careful descriptions of a designed solution to a recurring problem. The endorsement of parametric computation, in the first instance, allows generating more easily than before a core solution to design problems. It carries the overwhelming potential of using that solution many times without ever obtaining the same result twice. On the other hand, material transfers that associative linking in a sensorial realm, thanks to its physicality: in this domain, people can experience first-hand the space of architecture.

> The becoming form in the following presented design- and research-approach is thus always based in the possibilities and constraints of the actual materialization.
>
> (Menges 2009)

The seemingly abstract concepts introduced with Aristotelian thought begin to acquire an investigative value from an architectural and computational point of view.

1.3 The Virtual Value of Computing

1.3.1 Historical Report: Computational Rise

1.3.1.1 Parallel Between Design and Cybernetics

Beginning the search for some primary transfers of speculative statements in the computational architecture concerns, we definitely have to define computation as the process of calculating. Looking once again for lexical roots of what calculation means, we cannot disregard the minimum entity at the base of the calculation, i.e. the number or the digit its written sign.

di·git/'dɪdʒɪt/n [C] (1) one of the written signs that represent the numbers 0 to 9. (2) technical a finger or a toe

di·gi·tal/'dɪdʒɪtəl/adj (1) using a system in which information is represented in the form of changing electrical signals

(3) formal of the fingers and toes

(Longman Dictionary of Contemporary English)

Hence, according to common dictionary definition, we see that the digital commitment in architecture has to deal both with numbers and manual applications. For sure the most outright linkage consists in the use of automated calculus systems to obtain form and shapes. The mechanization of calculation, over the last century, has actually formed the theoretical foundation for the modern computing. Despite the rise of computation lays its basis since the Roman's abacus, its automated machine translation has occurred throughout cutting-edge inventions of adding machines[12] and mechanisms able to perform multiplication[13] developed during the seventeenth century or even the brilliant speculations[14] of the early nineteenth century. But what is worth to mention here are few keystones on which is based the birth of the epistemological dimension of cybernetics in architecture, both theoretical and practical. Mathematical logic and symbolic calculation paired with the pivotal introduction of electronics paved the way for algorithmic language,[15] contributing to reinterpret computing as a logical rather than arithmetical question. The innovative speed of electronics that overcomes the features of relay and switching circuits, brings to the invention of computer.

This new electronic machine takes advantage of the learning of analogic calculation but enables users to use digits to model a problem to be solved. The World War II played the role of a catalyst as its massive need for fast calculations.[16] If all the early applications of electronic calculators have served for military-oriented realizations, as happened also for the Cold War, the improved performances of Personal

---

[12]W. Schickhardt and B. Pascal started some pioneering work on calculating machines. Pascal is recognized as the first scientist to produce successful mechanical calculators: the so-called Pascal's calculators and later Pascalines.

[13]In 1673 the German polymath and philosopher G. W. von Leibniz designed a calculating machine called Step Reckoner. The machine expanded on Pascal's ideas, was able to perform multiplication by repeated addition and shifting.

[14]C. Babbage (1791–1871), along with A. Lovelace (1815–1852), is remembered for originating the concept of a programmable computer, even though the Analytical Engine mechanism remained uncompleted. C. Babbage designed a mechanical calculator that required a steam engine to operate, and that was too sophisticated to be built in his lifetime.

[15]See on this topic the pioneering work of A. Turing and his machine, dated 1936.

[16]"During the 50 years since the Second World War, a paradigm shift has taken place that should have profoundly affected architecture: this was the shift from the mechanical paradigm to the electronic one". These are the words that P. Eisenman uses to describe the advent of IT in architecture in his article Visions Unfolding: Architecture in the Age of Electronic Media already published in 1992.

Computer technology during the 1970s made possible its ubiquitous and effective spread. Besides the improvement of technological constituents of processing unit (e.g. magnetic core memories and mass storage, or input and output mechanisms), PCs mainly introduced a new vision of real: in particular human minds started to deal with complexity as result of interactions between simple elements. The analogy between human brain and computer, between neuron and bit of information, led the cyborg season. At that time, computer acquired the role of gigantic brains, the machines that think by E. C. Berkeley (1949), or even machines envisioning the cyborg season in literature and cinema.

But what are the influences of IT on design processes? Such interrogation became rich of interest for architects since when manifold and wide-ranging software, carrying *The Magic Inside the Machine*—as the Time's cover from April 1984 by D. Wynn (Computer Software 1984) clearly define computer software-, started to provide enriched features to "specialized elements, [...] connected by wires" (Weiser 1991). Undoubtedly, computer and graphic software have enhanced human investigation's capacities about formation, logic, geometry, system and algorithm. First and foremost, they provided the opportunity to deeply expand multidisciplinary and theoretical learning, giving shape to a CAD's agenda over the years. Therefore, it is possible to identify epistemological roots of digital in architecture, involving philosophy, logic, geometry and system theory that more or less consciously guided the architects' CAD practice.

We have to refer to the very founding seminal works of the Western system of thinking to fully understand the profound impact that computation have had on both the ideation and realization of architectural drawings and space. The pivotal thought of Aristotle has laid the foundations for the long-term evolution of the conception of form. The Holism (etymol. from Greek *hòlos*, meaning all, entire, or total) affirms that the whole is a system resulting from mutual interactions among constituents, and it is not the stark collections of individual parts. Several attempts in the field of art, literature or science have contributed to finding a theory, namely, morphology, bound to affect many disciplinary inquiries. It depicts the inferred character of a system by the complex link of form and formation, which result in translation from German words *gestalt* and *bildung*.[17] S. Kwinter, in his essay entitled *Who's Afraid of Formalism*, defines the matter of form as a long-lasting concern with the process of formation: the meaning that "manifest form" acquires as an ordering logic of patterns and embedded forces is opposing to the object as a simple final outcome (Kwinter 2007). This conception, which intends form as a method,[18] denotes a divergence between the outer shell of end products and the active processes that bring them into being, and it undoubtedly retraces J. W. Goethe's lessons.[19] The characters of behaviour (mostly natural ones) are decoded and conceptualized as iteration, gener-

---

[17]J. W. Goethe expresses these concepts in his botanical writings in the early 1800, introducing a science of morphology. Both from German, gestalt means form; while bildung means formation.

[18]It is considered appropriate once again to underline the etymology of the term. Method comes from the Greek *metà-hòdos*, literally meaning the way that takes you further.

[19]"Ernst Cassirer once said of Goethe that his work completed the transition from the generic view to the genetic view of organic nature. [...] Goethe's formalism, like all rigorous and interesting

ation and variation of geometric formation: these steps are ultimately fundamental to the execution of computation in design.

The mathematician D'Arcy W. Thompson, in early 1900, gives rise to a conceptual framework for formation (the German *bildung*) and transformation that can be borrowed from exact sciences and applied to manifold contexts. A parametric equation results in producing families of geometries where each instance shares a certain level of affinity with the others. Through these entities, which are homologous to each other, an individual course of development can be deciphered.[20] The study of form may be purely descriptive or it may follow an analytical path by implementing mathematical definitions; this level of symbolic abstraction meets soon the interest of architects. In fact, these theories bring the role of pattern to the architect's attention, though they are mainly related to natural growth processes.

Many reflections and experimentations in art, architecture and design after World War II were based on the assumption that information process gave birth to patterns that could be observed in natural and human behaviour. In considering design problems, C. Alexander established the "need for rationality": logics, mathematics and physics become suitable tools when

> used to explore the conceptual order and pattern which a problem presents to its designer.
>
> (Alexander 1964)

In particular, logic constitutes a convenient platform in matching the growing body of available information and it allows architects to take a substantial step forward: the conception of form shifts from a distinct object towards an externalized operating system. The growing of this system follows a "diagram of forces" (Thompson 1917) where a computing agent (the object) is distinguished from the computational process (the method). The proposed transfer of philosophical principles in architecture leads to the generation of emergent systems by identifying and combining a collection of informed behaviour.

On a theoretical standpoint, first episodes started to embed the theories of formation, transformation and growth of form within the design path and resulted instrumental to promote cybernetics concepts in the architectural investigative agenda. The "need for rationality" is then further developed within the virtual environment. The intellectual domain of cybernetics advances the research of the "goodness of fit" (Alexander 1964), by augmenting the transdisciplinary value of those systems previously introduced.[21] Cybernetics, defined "as a field that illuminates the concepts

---

ones, actually marks a turning away from the simple structure of end-products and toward the active, ever-changing processes that bring them into being" (Kwinter 2007).

[20]The Scottish polymath, in his book On Growth and Form, emphasizes the roles of physical laws as well as of mechanics in the determining process of the form in his most famous book. It covers many topics; here the only concern about Darwinian evolution is considered for the purposes of the narrative.

[21]In hazarding a parallel between the ages, the "goodness of fit" calls to mind the *concinnitas* (etymol. from Latin *concinnus*, meaning skilfully put together or join) adopted by L. B. Alberti. Inspired by the classical tradition of oratory and Cicero's rhetoric in particular, he emphasizes the aptness for or the adaptation for a purpose as the architectural form's topic.

of adaption and control by way of abstraction" (Riiber 2011), specifies that systems are built on regulation, adjustment and purpose, all informed by means of feedback.

The introduced "goodness of fit" can be defined as the cornerstone based on which is developed the Computer-Aided Design path[22]: CAD "is a cybernetic method and there are several instances of its application to architecture" (Pask 1969). The pioneering design and computing activity of I. Sutherland (Sutherland 1963) underlines the engagement of human constituent in the development of hardware and software supports. His pioneering man–machine system—namely Sketchpad—is the first effective accomplishment by which bits, as "DNA of information" (Negroponte 1995), are acting within a mechanism of adaptation and enabling control. In contrast to the perception of a building as merely static material object, the framework expressed by G. Pask promote the architectural activation of mutual and holistic systems previously only theoretically foreseen.

A peculiar feature of the dialogue between design and cybernetics arises from the genuine dialogue between forms and manifold disciplines (e.g. engineering, biology, economics): only a circular reasoning leads to adequate solutions and introduces the concept of an evolutionary architecture.

> It is precisely the ability of "finding a form" through dynamic, highly non-linear, indeterministic systems of organization that gives digital media a critical, generative capacity in design.
>
> (Kolarevic 2003)

The relentless computing power along with software development decidedly improves evolutionary ideas in architecture, where biology or physics no longer serve as simple inspirational sources, but are guiding principles for the back and forth design process.[23] When forms are no longer top-down designed but calculated they directly originate from an evolutionary model, which needs an 'architectural concept to be described in a form of "genetic code"' (Frazer 1995). Evolutionary algorithms do not simply follow instructions, but their codes run abstract solvers, continually producing new or unexpected outcomes (Kolarevic 2003).

The effective implementation in architecture of the foundational thoughts is undoubtedly dependent on the design apparatuses and their improvements. However, some fallacies impend when their design and development is solely intended as an instruments update and not as an ever-expanding toolkit for a methodological renewal.

---

[22]It may be schematically summarized as a climax started with a form-based phase or computer-aided drafting—where the architect still superimposes its final configuration and draws it with digital media, and reach its peak with the performance-driven-design—where the computerization moves towards a computational approach defining objective, performance and fitness criteria for a digitally developed design, passing by a form-finding design.

[23]"In a non-linear design process design directions and alternatives are generated, presented and evaluated simultaneously, and in real time" (Grobman et al. 2010).

1.3.1.2   The Default of the Virtual Domain

Although the use of CAD technologies as facilitative but supplementary means to accomplish free-form architectures may lead occasionally to innovative spatial qualities, 'it is important recognize that the technology used in this way provides a mere extension of well-researched and established design process' (Menges 2008). 'The digital era has added little value so far to the quality of design' (Frazer 1999) and the early attitudes of the so-called digital turn in architecture were a matter of inspiration and perhaps fascination. Thereafter, several investigations have driven a digital tectonics' exploration, involving computers generally as machines able to compute forms, though they already place at architect's disposal a versatile platform where innovate effective design methods. During the 1990s, 'the diffusion of complex forms continued on the one hand, sometimes condensed into stylistic derivations' (Barazzetta 2015). Some other times the inappropriate reliance on software has guided towards a misleading or rather a loss. 'The iterations of digital files, the native digital objects and data-sets, as well as the tools and machines used in their production are disappearing with every migration to a new operating system, [...] and every upgrade in hardware' (Lynn in Eisenman et al. 2013). Although the '(Great) Loss' referred to by G. Lynn in his book has purely practical consequences, it can be somewhat related to the 'mechanical loss of innocence' highlighted by C. Alexander (1964). This enterprising analogy emphasizes the relation between project and digital environment in which it is developed: by identifying the project idea with its tool, architect ousts spatial and material entities from the research. This attitude is 'characterized by the paradox of introducing the computer in architecture without actually introducing computation' (Thomas 2007).

The 'computational fallacy' (Kwinter 2003) basically arises from two oversights. On the one hand, there is the digital (or numerical, according to its Latin etymology, see footnote 11) over-abstraction indirectly connected to Terzidis' reasoning (2006). The Greek professional poses a valuable point in describing computation as not vitally bound to the computer. This frees it from the whole of geometrically and mathematically generated stylistic clichés and from the particular modes of software engagement (i.e. the interfaces) that has entrapped architects.[24] On the other hand, K. Frampton (1995), in the context of his studies in tectonics, considers the digitization[25] of architecture a threat to the physical aspects of construction. The '"dematerializa-

---

[24]"Today's digital tools [...] have been authored, as for example in the case of Bezier splines or Blinn shading, by engineers, mathematicians and hackers. In mistaking the creative work of the tool producer with that of the user-architect, the profession runs the risk of developing a discipline based on the consumption of the latest algorithmic technique or graphic software package and of having to keep pace with the developments of software fashion for designs to have 'the latest' geometrical flavour". These are the explanatory words by P. M. Carranza in the article Out of Control: The Media of Architecture, Cybernetics and Design published in 2007.

[25]The digital design process loses its initial aptitudes. "It appears that a large part of contemporary architecture is determined by algorithmically established design procedures in which the constructive and building implementation is of insufficient significance and appears secondary. Realizing a design, an image, or a drawing" was the project concern, while the architectural matter remains the materialization of space (Brell-Çokcan and Braumann 2013).

tion into pure form", as raised by digital architecture during the 1990s' (Gramazio et al. 2014a, b) got lost in the strongly deterministic nature of digital processes, without referring to construction or to materiality. Computer representations often tend to neglect the material and technological nature of architecture; or at least they confuse it with the restitution of a superficial cover through the procedure of texturing. 'Computer must play only the appropriate intermediary role of interface' (Kwinter 2003) and the unconditional acceptance of new computational tools should leave room to the resolving investigations of the materially indeterminate architectural visions.

Though the study of morphology has solid epistemological roots, its digital transfer in morphogenetic applications has revealed an 'emblematic one-dimensional reference' to theoretical notions resulting in shapes rather than in forms. Those shapes have remained 'elusive to material and construction logics' or at least they have left the matter to subsequent solutions that are pursued as top-down engineered material solutions (Menges 2008).

Thus, the post-digital to which the book refers is expounded by experimental projects, which are engaged "proactively in the creation and use of digital tools [both software and fabrication related] to reach otherwise inaccessible results" (Eisenman et al. 2013). *Gestalt* and *bildung* are the focus on which project is oriented not by complicated shape configurations,[26] but rather by complex spatial and material evidences.

Thinking back to the potential that Information Technology has introduced in architecture, we can assert that only recently computation has started to perform manifold tasks. Computer has been used only as an efficient multifunctional tool within a well-established design approach. Digital CAD, while depending on common ways of design, leads to complex blank geometries. Whereas outcomes of truly computational design approaches are characterized by the reciprocity of pattern and material, resulting in "an uncomplicated complexity" (Menges 2009).

Nowadays, computational design has strengthened procedures of organization, articulation, signification, prediction, simulation and evaluation of families of outcomes.[27] The computational stream enlists architects as cognitive agents, perceiving and decomposing systems of information. It allows either to deduce results from parameters or to simply weave complex sets of values (Menges and Ahlquist 2011). At the very beginning of the digital turn, Deleuze and Cache have defined the *objectile* as a pattern designed for interaction and variability. An *objectile*, as in the Aristotelian thinking, 'is a class or family of object, but no object in particular' (Lorenzo-Eiroa

---

[26]"The general tendency among many progressive practices developed around the 1990 s has been to explore and use the computer as a tool, but its effect on a discipline heavily structured around the media of production has seldom been addressed. This purely instrumental use of the computer has too often resulted in the production of intricate geometries [...], which are justified only by their mere possibility. Without the evaluation criteria the new conditions may require, such geometries are still treated and valued in relation to traditional architectural representations, [...] and at the same time inadvertently surrendering the understanding of the processes behind them" (Thomas 2007).

[27]In the last decade, many leading conferences started to debate and investigate these fields with many avant-garde case studies.

and Sprecher 2013). As if it were something that is not yet in place, but which has the potential to be able to become. Such form-in-potency, differently than a form-in-actuality, is seen as lacking compelling architectural features.

In this sense, the aforementioned etymological progenitor of form (i.e. pattern) underlines the intellectual fallacy pronounced by S. Kwinter.[28] Issues of representation and simulation but first and foremost of communication with non-experts are critical in addressing in a solely virtual domain. The open-ended pattern of links needs to break away from the bounds of possibility, claiming a material transfer. The latent potential arising from computational environment requires proper technology and tools to move from something that is not yet, to form that act. This compelling call for direct materialization is revolutionary and declares the default of the virtual domain. Rather than being only the necessary consequence of the tool's development, it demands a computational design chain intended for physical transformation. Within the digital background, the delayed conversion was undoubtedly due to methodical as well as technical obstacles.

Digital design and fabrication are strictly dependent from latest technology. Digging up the origins of the word in itself, we can have a look at what technology means. We easily find that the etymology is Greek: τέχνη was used to indicate the ability to outwit circumstances. Aristotle was the first to join the word téchne with logos (meaning word or reason or intellect) to form a single term téchnelogos. The fleeting appearance on his treatise *Rethoric* suggests a meaning as skill of words or speech about craft, but also activities involving the making of things guided by reason.

For the next thousand years, *téchnelogos* disappear as a word but technology of course not. Indeed throughout the centuries of the European Middle Age till the eighteenth century, with the overturning Industrial Revolution, arts and craft led significant machine developments. The civilization progress was made possible thanks to outstanding ideas, machine inventions and innovative uses of energy. *Téchnelogos*, and its synonym, reemerged thanks to the effort of the sixteenth century French P. Ramus, who used them in the more modern sense of "the *logos* of all relations among all *téchne*".[29] But in 1809, when J. Beckmann launched his book *Guide to Technology, or to the Knowledge of Crafts, Factories and Manufactories* that technology catalyses the interconnection between different branches of knowledge, arising as functional description of the production process. Summarizing in very few words, Beckmann's most important achievement was to recognize that human's innovations were not just a collection of fruitful inventions and profitable applications. *Téchne* and *logos* are newly linked within an inter-reliant sequence. Over the centuries and throughout its continuum of tooling developments specialized in disciplines, we can see that at the *téchnelogos*' core there are idea and information (Kelly 2011).

The intangible flows of information, which our mind is able to synthetize in results, is enormously amplified by the advent of computational culture. The intangible iden-

---

[28] Reference to Kwinter (2003).

[29] Reference to Szerszynski B (2005) *Nature, Technology, and the Sacred*, Blackwell Pub, Malden p 56.

tity of *téchnelogos* restates again that technic and words do not have a material body at all, as it is the case for the software. The architectural transfer of this ordered evolution cogently claims its material identity. The digital seamless path, accomplished by the file-to-factory technologies,[30] establishes a parallelism with the traditional relationship between project and craftsmanship, combining ideational procedures together with the rules of arts and crafts. The ability to digitally develop a three-dimensional model and then to use the same digital model seamlessly in the fabrication domain initiates the search of the demanded synthesis between design and construction.

1.3.2 Historical Report: Computational Fallacy

1.3.2.1 Added Value of Active Tools

Using Aristotle's concept of potency and actuality,[31] we describe the "new digital continuum" (Kolarevic 2003) as the way through which potential forms can act. On one hand, the possibility introduced by advanced software-houses allows to produce thousands of generation of optimized virtual constructions (that are forms-in-potency). On the other hand, the actual materialization of these latter is what is needed to fulfill post-digital architecture. Nowadays, digital tools (both hardware and software) have opened up the development of novel interfaces for interdisciplinary teams of developers.[32] The digital fabrication approach adds "value in computer-aided design" activating the "convergence of the virtual and the actual" (Frazer 1999).

It was only a decade after the pervasive use of CAD, when designers started a fruitful avenue of research around the full-scale materialization of architecture, challenging new CAD/CAM technologies. They follow Mitchell's advice. He has stated: "[…] through interfacing production machinery with computer graphic systems a very sophisticated design/production facility can be developed".[33] As it has already happened with computer logic, first instances of digital manufacture were relegated to a technical task, and moreover at the scale of models adding "relatively little to the design process" (Frazer 1999). Not long after, digital fabrication facilities became active media and actuators, instead of remaining passive and awkward mechanical tools.

We define the digital fabrication as a platform first and foremost of methods, strategies and therefore also of tools introducing a follow-up to the jammed computational

---

[30]Here the direct reference is to CAD and CAM technologies; in particular to their inter-communication allowing the physical implementation of computed shapes.

[31]See Aristotle, *Metaphysics*, translated by W. D. Ross.

[32]Architects, artists, biologists, engineers, and so on can share their expertise and effectively collaborate together.

[33]Reference to Mitchell, W. J. (1977). *Computer-aided Architectural Design*. New York: Van Nostrand Reinhold.

scenario. It provides tools, not only instruments, to actualize the direct transfer from computational thinking to digital manufacturing. But it does that in an active way in which material plays a predominant parametric role. Digital manufacture techniques empower design and encourage to revisit materials and reconceive their manipulation processes by incorporating them as variables in the computational continuum. Even fabrication media are considered as parameters from the very initial stages of design. Fundamental in this perspective is the evolution of CNC machines. Borrowed from industrial field, apparently far apart architecture, they start to be adopted and adapted by seminal academies and worthy design practices. When Kieran and Timberlake (2004) express the need for transferring technologies, they are encouraging architects to adopt output-based production techniques from other industries in order to investigate material as well as tool behaviour. The resulting range of unpredictability constitutes the field from which innovative design solutions may emerge. The controlled percentage of errors granted by tooling represents the fissure whereby the global design demonstrates the added value of active involvement of tools.

During the 1990s, in the fertile context of technological advancements, B. Streich, at the Department of CAAD and Planning Methods (University of Kaiserslautern) focused his research[34] on CAD techniques applied to architectural models. The Semper Pavilion presented at Archilab in Orléans (France) was the first example where the seamless digital chain has taken place.[35] From design procedures to manufacturing process, everything was generated within the same virtual engine. A CNC router received lines directly from designer's computer, without any additional control by any third party. Every detail was algorithmically and automatically designed and manufactured and the final wooden interlacing surfaces refer in an iconic way to design that is intertwined with fabrication.[36]

In building domain, F. Gehry has had a pivotal role: digital information starts to drive CNC production without producing drawings. The Golden Fish (1992) in Barcelona is recognized as the first project using advanced modelling techniques in the field of building construction. The potential shape was modelled within the virtual environment of Autodesk Maya, and subsequently structurally optimized with CATIA by Dassault Systèmes. This digital model fully directed the production and assembly on-site of each component putting into practice the flow from forms-in-potency to forms-in-actuality.

At an earlier stage of the digital turn, forms have broken their traditional Cartesian guise fixed on sheets of paper or into screen. Now, they start an effective process of extrication from their potential aura. Designers while defining behaviours, rather than fixed dimensions, start to evaluate the fabrication process as an informing parameter.

[34]Reference to Streich B, Weisgerber W (1996) Computergestützter Architekturmodellbau: CAAD-Grundlagen, Verfahren, Beispiele, Birkhäuser, Basel, Boston.

[35]Information retrieved from http://architettura.it/image/festival/2002/en/texts/cache.htm [last accessed January 17, 2018].

[36]The issue of AD guest-edited by M. Taylor was significantly titled *Surface Consciousness*. Reference to Cache B (2003) Philibert de L'Orme Pavilion: Towards an Associative Architecture Architectural Design 73:20–25.

The world of bits[37] converge towards material actuation; studying how to "turn data into things" (Gershenfeld 2012) was the aim of the Center for Bits and Atoms established at the Media Lab ran by professor N. Gershenfeld (MIT). Since 2001, the research was focused on how merge computer science and physical science, two fields corresponding to software and fabrication domains.

The opportunities emerging from the availability of digital fabrication tools lead the development of an increasingly solid interface area between forms-in-potency and forms-in-actuality. As an ecotone is a transition area between two biomes in biology, here manifold expertise have the possibility to act within a whole process file-to-construction thanks to digital fabrication.

In this unifying vision, the architect is conceived as the master builder, likewise it happened before the Renaissance: he is the main figure who joins ideation and construction as two links of a continuous chain. This interface area is "opening up radical new opportunities for the practice of architecture" (Sheil 2005), "defining how to realize elements given a certain (algorithmic) description, and how to synthetically describe the physical environment and behaviour using our computers".[38]

1.3.2.2 Mass Customized Material Accomplishment

A crucial consequence of the so-called third industrial revolution in architecture has certainly been a strong advancement in tools. Alongside the mass production as undeniable achievement, it has triggered "advanced customization which supersedes the resilience to innovation typical for the conservative sector of building construction" (Barazzetta 2015). In architecture and in building industry, the advent of rationalized production processes accomplishes serial unique elements. The result of industrialized methods borrowed from the automotive compels the use of repetitive construction elements in order to be cost-efficient. The decidedly non-traditional algorithmic forms break the serial production because restrictive and not as effective as once thought.

The new options offered by CNC machines prompts architect to customize their fabrication tools. The relentless attitude of actively designing bespoke building component, material strategy as well as digital manufacturing tool is known as mass customization.[39] S. Davies[40] has coined the term, meaning more than an intriguing

---

[37]This trend is epitomized by enlightening works as *Being Digital* by N. Negroponte (Negroponte 1995).

[38]Reference to Katz R (2002) Bits and Atoms: An Interview with Neil Gershenfeld. *EDUCAUSE Review Magazine*, 37(2). Retrieved from https://er.educause.edu/articles/2002/3/educause-review-magazine-volume-37-number-2-marchapril-2002 [last accessed January 17, 2018].

[39]From the Wikipedia's entry, it is the use of flexible computer-aided manufacturing systems to produce custom output. Those systems combine the low unit costs of mass production processes with the flexibility of individual customization. Retrieved from https://en.wikipedia.org/wiki/Mass_customization[last accessed February 11, 2018].

[40]In his book, S. M. Davis supplied both a name and a conceptual framework for processes then taking hold in the clothing industry. He recognized that mass customizing simply extended the

oxymoron: in exploring innovative business and organizational possibilities, he provides a conceptual framework for a then emergent process. The new post-Fordian
paradigm for emerging forms of production[41] is contextualized within a revolutionary
industrial change driven by a novel approach in the production of architecture.

It has begun a rapid shift towards new manufacturing scenarios in architecture
involving innovative features. We can briefly list the personalization capacity originating by computational versatility and vicissitude,[42] the high-level of precision and
repeatability (coming from the CNC tool adaptation), and a truly effective design-
to-production chain. The architect's figure emerges as a new identity demanding
manifold skills spanning from technical coding to analogic material manipulation,
from speculative to experimental aptitude.

Initially, mass customization in architecture has remained a localized promise and
less systematic[43] if compared to industries with high production volumes. Nowadays, it represents a first step towards an individualized fabrication, which matches
the demand of parametric design in producing unique outcomes. Kieran and Timberlake (2004) provide a conclusive definition: while investigating the refabrication
process of architecture, they interpret the mass customized production as a process
capable to build using automated manufacture, but with the ability to differentiate
each artefact from any other fabricated during the same chain. These words call back
the opinion that M. McLuhan expressed in 1969: "once electronically controlled
production has been perfected, it will be practically just as easy and affordable to
produce a million different objects as to create a million copies of the same object".[44]
CAD/CAM technology reduces the need for repetition of standard elements, increasing the feasibility level of repetitive but not identical components, distinctive results
of computed forms. The digitally driven production processes introduce a different
logic of serialism in architecture, one that is based on local variation and differentiation in series (Kolarevic 2003). Variation is not only of the final products, but also
first and foremost of inventive and fabrication procedures.

The first step towards such tailored fabrication looks at expanding competi-
tiveness by reducing waste of resources, time and material above all. And in fact
customization is relevant not only regarding the final outcome, but first and foremost when it is embedded within the digital design workflow considering the self-
construction of building equipment as well. In fact, one of the main benefits is the
concrete possibility to produce bespoke machine effectors customizing not only software environments.

---

capabilities latent in all CAD/CAM processes. Reference to Davis SM (1989) *Future Perfect*.
Reading, Addison-Wesley Publishing Massachusetts.

[41] B. J. Pine II, in expanding Davis's definition, separated production into three categories: craft
production, mass production, and mass customization; this latter combines elements of the first two.
Reference to Pine II (1992), pp 50–52.

[42] It refers to the inspiring issue of AD 78(2). Hensel M, Menges A (2008) *Versatility and Vicissitude:
Performance in Morpho-ecological Design*, Wiley, London.

[43] This description is found in Pine II (1992).

[44] This statement was meant to be pervasive. Reference to McLuhan M (1969) *Mutations 1990*,
Hurtubise HMH, Montréal.

1.3.2.3 Digital Materialism: A Peer-to-Peer Process

It is undeniable that digital means of production are becoming more and more accessible, following what has already happened for computer and software in the late 1990s. But above everything, the gradual acceptance of cutting-edge digital manufacturing media encourages designers to rediscover architecture as a material-based practice. This shift away from the representational modes of forms implies that buildings and architectural objects regain their vanished material identity. Historic architectural process, "in which the role of master builder was assumed" has been progressively replaced by fragmented specialization "in which architects have diminished the importance of their own involvement in the act of construction" (Booth 2009). Conversely, the adoption of CNC fabrication machines has enhanced the relationship between architect and new tailored materializations *en masse*.

Different matters turn into material when the human activity of making engages their singular attributes as design parameters (Kanaani and Kopec 2016). Then, digital materialism[45] is a structural system in which space-defining, load-bearing and energy-conducting elements arise from a computational set of information transferring also matter properties. In a peer-to-peer process the very materiality and its physical features, the constraints and the logic of design, production and assembly processes constitute the means of expression of research projects.[46] In other words, architects communicate the means through which their forms came into fruition, setting up a system of work providing a material expression to computational structure. Although a technical involvement with machines is needed, in architecture an intensive material implementation is compelling (Gramazio et al. 2014b). Only questioning the very process by which architects fabricate architectural design, we can go beyond the shortcomings of efficient but mere technical ameliorations. The inbuilt understanding of material and fabrication variables, along with computational capacity, empowers new modes of digital materialism. As N. Leach clearly recognizes the "emphasis is therefore on material […] over appearance, and on processes over representation".[47] The focus moved really far away from the top-down approaches in which the objective was simply to employ computer to make the architect's ideas reachable.

Gramazio and Kohler's motto "digital materiality" (Gramazio et al. 2008) initiates a popular trend. In the early 2000s, their group at ETH in Zurich has started laying foundation for a fascinating and resourceful research line, where computing power acquires a renewed role. The computational chain is enriched by the new

---

[45]Philosopher M. De Landa introduces his "new materialism" as novel understanding coherently articulated with science and technology. This new conceptual framework has several historic reference points: some are improved while others are totally transformed. Firstly, the Aristotelian view of material as passive recipient of predefined or more precisely predesigned shape is gone. Material can act as an effective agency in design.

[46]The definition of material systems within the computational scenario is well expressed by A. Menges. Reference to Menges, A. (2012). *Material Computation: Higher Integration in Morphogenetic Design*. Hoboken, N.J.; Chichester: John Wiley & Sons.

[47]Reference to Leach, N. (2009). Digital Morphogenesis. *Architectural Design,* 79(1), p. 34.

link of robotic fabrication that effectively bridges the gap between form-in-potency and form-in-actuality. Besides the unprecedented technological evolution of tools, in here the most important revolution is challenging the deep-rooted hierarchical approach: "geometric information being prioritized over its subsequent materialization" (Menges 2012). Rapidly the body of work on the "manifestation of data" (Burry and Burry 2016), particularly developed by academic institutions, has undertaken researches at a higher level of thinking and execution.

"This context might very well lead to a redefinition of the professional identity of the architect, besides modifying the nature of his production" (Picon 2010). Given that, new digital fabrication techniques are expected to broaden design scope, in the quest for a new design paradigm based on digital design and production, next section investigates new fabrication availability and processes, above all.

1.4 Man–Machine Architecture

The Machine is the architect's tool - whether he likes it or not.

Unless he masters it, the Machine has mastered him.

(Frank Lloyd Wright, 1908)

Manufacture in architecture takes on special meaning since it refers to a wider objective than the representational and abstract intentions behind modelling. Digital manufacturing has newly reawakened prototyping in architecture: it establishes an enlightening process through which designers gain understanding into how experimentations proceed. Even failures offer important information; when fallacies are reintegrated within the computational process as part of the feedback loop, they induce changes broadening the boundaries within which the prototype is sometimes forced. However, effectual or unsuccessful projects lead to unexpected consequences in terms of opening new fields of research.

The aim of this sub-chapter is to offer an overview of the digital fabrication network. Informed CNC manufacturing processes seamlessly link 3D geometry to its final materialization bypassing the production of drawings. The digital continuum (Kolarevic 2003) improves accuracy and "makes complexity ordinary" (Tedeschi and Wirz 2014). The dissertation is intended not as machine instructions but it critically questions the extent to which hardware tools are affecting the whole architectural design process.

The resulting awareness encourages considering the digital fabrication approach as a design parameter, involving material factor as well. In terms of design and construction of any architectural investigation, it is fundamental to understand in advance which techniques can be adopted and the specific machine typology to be used for its fabrication, in order to integrate them as design variables and limitations as early as possible. In this section, the emphasis on techniques for prototyping refers to the impact of CNC technologies in architecture especially questioning how fabrication aspects are fed back into the design process. Those task-specific processes are then enhanced by avant-garde robotic implementations. The following paragraphs explore additive, subtractive, transformative and assembly technologies.

1.4.1 Machining Process or Computational Coalescence

Looking back over the early twentieth century, machine-aided fabrication systems of modernism has represented the Holy Grail of architecture: a means of exerting greater control over the design process by being cost-effective and reducing the obstacles of conventional fabrication process (Menges 2012). By the late 1990s, the diffusion of CAM activities within the worldwide academic and practice fields has triggered a search for the architectural relevance of digital fabrication accomplishments. The advent of a favourable practice, arising from "techno-enthusiasm", demands an in-depth immersion in the digital fabrication's matter. Renowned 3D printing seems to evoke a jobless technology, as one front cover by The Economist[48] illustrates. As already happened with previous technologically driven seismic shifts, CNC machining process is recognized as the emerging core of design practice, instead of being a barren process (Sheil 2012).

So what's relevant is not a competing rivalry on the size of laser cutting and milling machines, or robot-cell's equipment, or even custom-engineered end-effectors availability. Rather an effective computational coalescence[49] is crucial, where architecture reorganizes as an experimental-investigation by orchestrating virtual dataset with the physical object itself.[50] Because of that designers lead a concerted process, not solely machining operations. However, the possibility to gain more insight of the strengths and weaknesses of computational architecture cannot be conceived without *téchnelogos* learning.

The vast scenario of digital fabrication set-up for architectural projects includes several digitally driven manufacturing techniques adopted from other industrial fields but also and particularly it is a net in continuous extension. Since 1952, when MIT entered the market the first XY cutter, the CNC revolution has begun. At its most basic CNC devices consist in a tool precisely moved in the three dimensions. The control is performed by stepper motors, which differ from the brushed direct-current motor because they convert a series of input pulses into a precisely defined incre-

---

[48]Reference to Print Me a Stradivarius. (2011, February 12). *The Economist, 398*(8720). Retrieved from http://www.economist.com/node/18114327 [last accessed January 17, 2018].

[49]As the dictionary entry suggests, this coalescence is literally a process where computational design combine objects and ideas in one single hands-on approach. Coalesce. (n.d.). In *Longman Dictionary of Contemporary English*. Retrieved from http://www.ldoceonline.com/dictionary/coale sce [last accessed January 17, 2018]. Reference to Menges A, Coalescence of Machine and Material Computation. In: Lorenzo-Eiroa and Sprecher (2013), pp 275–283.

[50]Architecture is an investigation "partly a computational orchestration of robotic material production and partly a generative, kinematic sculpting of space" as P. Zellner has observed in his work *Hybrid Space*. Reference to Zellner P (1999) *Hybrid Space New Forms in Digital Architecture*, Rizzoli, New York.

ment in the shaft position. By using computer-generated pulses much finer movement can be achieved by determining number of steps in a rotation and the speed
as well. Nowadays, the increasing sophistication in controlling CNC machines goes
beyond the number of axes per machine (usually three- or five-axis machines are
the most frequent) or the precision of stepper motors. It refers to software interface
and feedback loop allowing the ability to interact with and adjust the programmed
stream of information according to uncontrollable factors, e.g. material variations.
Furthermore, highly sophistication might enable CNC devices to make real-time
adjustments receiving information from sensors, e.g. anti-collision system. All possible refinements drive towards an automation (Andia and Spiegelhalter 2015) of
the fabrication process in architecture, fully settled in some academic robotic laboratories. Prior to automation, "wasteful and vulnerable channels that pass through
consultants, contractor, specialist contractor, maker and supplier" (Sheil 2012) usually have characterized the construction. Such scenario of advanced fabrication set-up
has the consequence of involving architects in every step of the digital continuum,
getting involved themselves with the means of production and their coding.

Architects enthusiastically adopt CNC routers, laser cutters and 3D printers in all
facets, bringing into relief the activity of prototyping. If, on the one hand, detractors
assert that with digital prototyping has disappeared the deep interface with detailing
and assembly experienced by hand workers, giving an illusion of making, on the
other hand, there is an undeniable cognitive potential. CNC processes have proved
their aptitude as affordable ways performing the new digital continuum (Kolarevic
2003); at the same time, they leave that unexpected variation still occurs during
the materialization of phenotypes from a genetic archetype (Burry and Burry 2016).
What is new of digital fabrication facilities is the motion control of the operating tools
along a path no longer guided by a technician[51] (Corser 2012). Digital fabrication
facilities offer architects the considerable possibility to customize own fabrication
tools, considering them as powerful drivers for the multidisciplinary design team. The
advent of a robotic season within the academic research demonstrates this attitude.

In fact, nowadays, the robotic fabrication generation in architecture looks at
the design and production of customized tools, almost three decades after the first
Japanese efforts for the application of robotic arms at the building scale.[52] CNC
machines gradually are brought out of the controlled industrial environment and
their production logic informs the designing path. Even though digital fabrication

---

[51]The aforementioned interface between the craftsman's hand and its artefact occurs by the neu-
rological and muscular feedback loop of a human technician, thus some inaccuracies may appear.
R. Corser states in his book that the primary tools of digital fabrication appliances are archaic. He
supports this statement with some examples about the subtractive process, described in a following
sub-chapter: he compares the laser cutting technology and the rapidly rotating router bit as refined
application of the "partial rotary motion tools, such as the fire drill, bow drill, and pump drill used
as boring and cutting tools throughout the archaic world". Reference to Corser (2012), p 154.

[52]During the 1980s large Japanese contractors have developed robotic production set-up in order to
improve working conditions and increase productivity. Reference to Hisatomi Y (1990) Introduction
of construction robotics in Japan. *IABSE Journal* 4 pp 26–30. ETH, Zurich. Retrieved from http://
www.e-periodica.ch/digbib/view?var=true&pid=bse-pe-003:1990:14::5#11 [last accessed January
17, 2018].

machines are able to manipulate different materials, they can be sorted into common broad groups. Clusters of additive, subtractive, transformative and assembly processes are set according to procedures that machines perform. While additive, subtractive and transformative processes are three common and widely recognized categories of local production, the assembly belongs to a different global sort: it is referred to the possibility of set-up proper production chain (from rough material to finished product) for non-industrial robotic environments.

1.4.2 The Robotic Arm as Architectural Enabler

While engineers primarily executed the early research on robotics, the real innovativeness of robotic fabrication arises only when architect exploration's objective starts to involve the forgotten multifunctionality[53] kernel of industrial robot as a design factor. Instead of colossal mechanical monsters, such as the O. Patent[54] or utopian proposals like Archigram's Walking City,[55] and highly specialized robots, the current focus is on "architectural robotics" (SuperLab 2016). The simple opportunistic application of robots in construction industry has not changed the design conditions of architecture. Over the last two decades, the indispensable conceptual leap has arisen from architecture schools: taking advantage from general research into industrial robots, which has involved mechanical and electrical engineers, as well as computer scientists and mathematicians since the 1950s, architects have started to reuse industrial robots and to adapt them by developing custom software interfaces and end-effectors. Industrial robots in architecture are compellingly implementing "computation applied to objects" (SuperLab 2016).

The level of abstraction, as the mathematical and logic procedures introduced since the advent of CAD tools, should drive the "robotic touch" (Gramazio et al. 2014b) to claim its own background. Robotic arms are playing an important role in architectural research, as their working space, low price, and inherent multifunctionality make them very suitable for architectural applications. Their current impact should not force the product's identity to cybernetic fascination coming from mere extensions of the designer's hands and mind.

Pondering the true nature of computation, robotic technologies and material science allow us to consider their mutual potential in encouraging the finding of an

---

[53]The ISO Standard 8373 defines an industrial robot as a multipurpose manipulator. Retrieved from http://www.iso.org/iso/catalogue_detail.htm?csnumber=15532 [last accessed January 17, 2018].

[54]The Demon Which is Destroying the People, an undated anonymous political cartoon, shows Oliver Patent—Monopoly in Crucible Steel—as a giant steam driven robot rampaging through a scattering crowd of people. This tragic representation epitomizes the aforementioned techno-pessimism. Reference to Fok WW, Picon A (eds) (2016) *Digital Property Open-Source Architecture*, Wiley, London, p 98.

[55]See http://www.archigram.net/projects_pages/walking_city.html [last accessed January 17, 2018].

emergent balance. Operating within this ecotone[56] architects blaze a trail for a new conceptual approach that differs from the endeavour of the past. Rather than focusing on the technological development of robotics itself, it is important that architectural conditions associated to materialization inform the approach to robotic fabrication, and not vice versa (Sheil 2012). One more time, the refinement of material, technological and epistemological values as well as aesthetic expression encourages designers to think differently and allows unprecedented levels of effective and direct collaboration across fields of expertise. Applying robots for nonstandard manufacturing remains unusual, but for building industry it is a necessity: architects only by "persistent efforts" (Brell-Çokcan and Braumann 2013) can establish and master a robotic process as a means of control of the continuity between the physical and the electronic, the tangible and the virtual (Fok and Picon 2016). Robotics, coupled with computational methodology, is challenging the scholastic perception: "the complex and the multiple appear more and more as the natural condition from which designers should start" (SuperLab 2016).

Facing robotic fabrication in architecture does not simply mean relying some lines of script in order to compute the G-code directly from the digital model. Robotics in architecture, first and foremost, unleashes methodical adaptability throughout the whole file-to-factory process. This goes beyond the "maker perspective" (Anderson 2012): in fact, the robot is no longer merely a mechanical sub-worker, but provides to architects distinctive wherewithal in order to design working prototypes, advantageously competing with the industrial mass production. Apart from that, a minimum knowledge of the logic and mechanics that lurk behind the robotic fabrication methodologies is required.

Articulated robotic arms require appropriate conditions, design strategies, kinematics, programming and control. A brief technical introduction is crucial in order to understand industrial robotic arms as accessible ideational members. In this way, their inherent performative potential can be accomplished not only during an experimental phase of research and prototyping[57] but also during the construction process moving away from the imitation of long-established construction technologies (SuperLab 2016).

1.4.2.1   History and Value of Robot

Although the concept of automata dates back to myths of Crete and ancient Greek talents,[58] the term robot was the brainchild of the Czech playwright, novelist and

---

[56]Refer to the biological comparison stated in the paragraph *Added Value Of Active Tools*.

[57]All robotic fabrication plants in academies are developing totally new construction method according to tailored end-effectors, techniques and materials that are involved.

[58]See the work of the Greek engineer Ctesibius or Archytas, previously cited as the progressive founder of mathematical mechanics. Reference to Rosheim M E (1994) *Robot Evolution the Development of Anthrobotics*, Wiley, New York.

journalist K. Čapek,[59] and introduced in his science fiction play *Rossum's Universal Robots.*[60] The Slav writer has envisioned a futuristic factory that assembles artificial people to carry out mental tasks. His view embeds robot into a cultural milieu: in confronting industrialization in a radical way, he proposes an integrative cooperation between human and machine. Čapek thus laid the basis for a pre-digital definition of the machine that is not merely mechanistic (Gramazio et al. 2014b). The original Czech names *laboři* and *robota* can be translated as forced labour: in the theatre piece, humanoids are deployed as cheap workers who rebel and destroy the humankind.[61] Consequently, the common sense, pairing work and servitude identifies robot as an automatically operated machine that replaces human effort.

The popular culture's viewpoint on robot is predominantly influenced by M. Shelley's book[62] and I. Asimov's nine short stories.[63] But also humanoid characters, such as C-3PO and Robocop, have had impact on popular culture. Besides these robots, real-life devices, such as Atlas[64] or Motoman,[65] have reinvented the industrial workplace and collaboration. They can be adopted as case studies for describing the components of robots or the complicating factors that make robotics so challenging (Jordan 2016).

As a consequence of this background, though not humanoid in form, multipurpose manipulators with flexible behaviour, automatically controlled, and reprogrammable have been developed for industry. Since its invention occurred in the first half of the 1950s, the industrial robot has started to pave the way for the upcoming automation applications. According to its ISO definition[66] and G. Devol's patent[67] perform repeated tasks with greater precision and productivity than a human worker. The

---

[59]He didn't coin the term: his brother, the painter and writer J. Čapek, is the actual originator. The original idea was to call the contraption *laboři* (from Latin *labour*, meaning worker). For more details see https://en.wikipedia.org/wiki/Karel_%C4%8Capek [last accessed January 17, 2018].

[60]Reference to Čapek K, Selver P (1923) *R.U.R. (Rossum's Universal Robots): a fantastic melodrama.* Garden City, N.Y.: Doubleday, Page.

[61]The vision of mechanical threatening monsters finds several paradigmatic exemplars with a political relevance during the 19th and 20th centuries.

[62]Reference to Shelley M W (1818) *Frankenstein, or, The Modern Prometheus,* London, Printed for Lackington, Hughes, Harding, Mayor & Jones.

[63]The stories, originally appeared in the American magazine Super Science Stories, were then compiled into a book. Reference to Asimov I (1950) *I, Robot,* Gnome Press, New York.

[64]Retrieved from http://www.bostondynamics.com/robot_Atlas.html [last accessed January 17, 2018].

[65]Retrieved from https://www.motoman.com/about [last accessed January 17, 2018].

[66]An automatically controlled, reprogrammable, multipurpose manipulator programmable in three or more axes, which may be either fixed in place or mobile for use in industrial automation applications. ISO Standard 8373: Manipulating industrial robots—Vocabulary. The definition refers mainly to machines for specific tasks repetitive, hazardous, and exhausting when manually accomplished. Retrieved from http://www.iso.org/iso/catalogue_detail.htm?csnumber=15532 [last accessed January 17, 2017].

[67]Together with J. Engelberger, the American inventor applied for a patent on Programmed Article Transfer that introduced the concept of Universal Automation or Unimation. U.S. Patent 2988237A was issued in 1961.

company Unimation Inc. initially developed material handling and welding robots. They installed their first industrial robot Unimate on General Motors assembly line in 1961. It was used to extract parts from a die-casting machine and to stack them. Other automotive industries, such as Ford Motor Company, purchased robot and started to use Unimate to spot welding as repetitive operations.[68]

The industrial robots of the first generation were conceived to replace human labour and the dexterity of arms. Robots capably repeat recurring operations as main assignment, but this is also their restriction. In this, they were not corresponding to the previously envisioned sophisticate humanoids. The industrial robotic arm with six degrees of freedom (DOF), which is used in almost every industry application today, was developed during the 1970s.[69] Over that decade, several industrial brands have produced specimens of articulated robots[70] marking the end of the era of hydraulic-driven robots (Naboni and Paoletti 2015). By the mid-1980s, the economic boom in automotive—particularly prosperous in Japanese automotive sector—has sanctioned the robot as a common industrial resource (Sheil 2012). At that time-building industry has started to implement robotic systems specifically oriented towards automation and prefabrication. In fact, in 1985, Kuka introduces a new Z-shaped robot arm whose design ignores the traditional parallelogram. It achieves total flexibility with three-translational and three-rotational movements for a total of six DOF. The new configuration saved floor space in manufacturing settings.[71] These first efforts were conceived to perform specific tasks, such as smoothing concrete, assembling form-work, and painting. They all were focused on monetary efficiency and reducing human labour achieving an increased level of final production.[72]

Although many technical improvements and mechatronic perfections have occurred during the 1980s, the industrial robot has been employed in building indus-

---

[68] In-depth analysis can be found in O'Regan G (2013) George Devol. In: *Giants of Computing*, pp 99–101. Springer London. Retrieved from http://link.springer.com/chapter/10.1007/978-1-4471-5 340-5_21 [last accessed January 17, 2018].

[69] In 1973 KUKA moves from using Unimate robots to developing their own Famulus: it was the first robot to have six electromechanically driven axes. In 1983 at pioneering robot company Unimation, V. Scheinman invented the Stanford Arm, the first all-electric computer-controlled 6-axes mechanical manipulator for assembly and automation. Scheinman commercialized the robot arm as the PUMA, or Programmable Universal Machine for Assembly, as we know it today. Reference to http://www.ifr.org/uploads/media/History_of_Industrial_Robots_online_brochure_by_I FR_2012.pdf and https://en.wikipedia.org/wiki/Programmable_Universal_Machine_for_Assembl y and https://en.wikipedia.org/wiki/Victor_Scheinman [last accessed January 17, 2017].

[70] Different brands, as ASEA or KUKA, presented their microcomputer-controlled all-electric indus-trial robot with six degrees of freedom able to perform continuous path motion. Reference to Caneparo (2014) and to Siciliano and Khatib (2008) *Springer Handbook of Robotics*, Springer, Berlin.

[71] Information retrieved from http://www.ifr.org/uploads/media/History_of_Industrial_Robots_onl ine_brochure_by_IFR_2012.pdf [last accessed January 17, 2018].

[72] For more detailed analysis see Bock T, Linner T (2015) Robot-Oriented Design Design and Management Tools for the Deployment of Automation and Robotics in Construction, Cambridge University Press, New York.

try in the same way of the origins: it has remained an efficient automated manipulator which repetitively fulfils only one specific task.

When talking about robotic fabrication in architecture we refer specifically to articulated industrial robots. Among all others features, the fundamental property attractive for architects is the versatility. Indeed, a robotic arm is suitable for differentiated tasks because it is not constructed for any specific application; rather initially it is not process specific,[73] similarly to the computer (Gramazio et al. 2014b).

In general, robotic arms can be catalogued in seven different types: gantry robot,[74] cylindrical robot,[75] spherical robot,[76] SCARA robot,[77] articulated robot,[78] parallel robot,[79] and anthropomorphic robot[80] (Caneparo 2014). They all consist of a mechanical arm and a numerical control unit. A typical industrial robot consists of an arm, a control cabinet, a teach pendant, and interchangeable end-effectors; this latter is the only specific task-oriented part of the robot. The tool is often custom developed by each design team and it is targeted to the material manipulation and constructive process that has to be performed. Many other peripheral equipment mounted on external axes are part of the robotic cell as well. The arm in itself presents a series of rigid metal segments connected together by joints or axes in a kinematic chain. The joints[81] allow the mechanical arm to carry out rapid and highly precise movements, reaching freely any points of the work envelope[82] in a nearly infinite combination of rotations.

To a large extent, the physical construction of an Z-shaped robot resembles a human arm; each of the six axes corresponds to a part of the anatomy: starting from

---

[73] Articulated robotic arm constitutes a generic piece of hardware that only becomes specific through the control software and the effector.

[74] It is also called Cartesian coordinate robot: it has three prismatic joints, whose axes are coincident with a Cartesian coordinator. Retrieved from https://en.wikipedia.org/wiki/Cartesian_coordinate_robot and http://www.allonrobots.com/cartesian-robots.html [last accessed January 17, 2018].

[75] It is a robot whose axes form a cylindrical coordinate system. Retrieved from http://www.allonrobots.com/cylindrical-robot.html [last accessed January 17, 2018].

[76] It is a robot whose axes form a polar coordinate system, in fact it is also named polar robot. Retrieved from https://en.wikipedia.org/wiki/Spherical_robot and http://www.allonrobots.com/spherical-robots.html [last accessed January 17, 2018].

[77] This robot has two parallel rotary joints and performs planar movements. Retrieved from https://en.wikipedia.org/wiki/SCARA [last accessed January 17, 2018].

[78] This robot arm has at least three up to six rotary joints; it is usually paired with external axes such as rotary table or linear rail. Retrieved from https://en.wikipedia.org/wiki/Articulated_robot [last accessed January 17, 2018].

[79] It is also known as generalized Stewart platforms: the end effector of this arm is connected to its base by three or six independent linkages working in parallel. Retrieved from https://en.wikipedia.org/wiki/Parallel_manipulator [last accessed January 17, 2018].

[80] It's a robot with its body shape built to resemble the human body. It is also named humanoid robot. Retrieved from https://en.wikipedia.org/wiki/Humanoid_robot [last accessed January 17, 2018].

[81] There are two types of joints: telescopic, which allows linear movement between two segments; and rotary (or hinged), which only allows revolving movement, in a single plane, of one segment compared with the other.

[82] Also named robot reach.

the bottom the base corresponds to the foot, the second joint is the waist, the third is the shoulder, the forth is the elbow, the fifth is the wrist pitch and the sixth coincides with the hand.

1.4.2.2   The Digital Master Builder

Visualize a construction site with its security fence system, in a hilly area of a developing French town. The land is levelled and ready to host a new building. Materials are stocked on the plot and the construction is already begun in the far corner of the frame. This is the scene that Villemard reproduced in his *Chantier de Construction Électrique* dated back in 1910. But this chromolithography—belonging to the collection *Visions de l'an 2000*—offers interesting traits to consider within our storytelling. It foresees a possible robotic construction site; in fact beyond ordinary elements appearing in any building site, what is strange to see (at least at the beginning of the last century) is a big machine operated by a single worker. He is sitting inside a cabin, identified by the word *architecte*: he is the designer who is actually and directly building his ideas. He is consulting the project (a sheet with the printed word plan) and manoeuvring through a panel of switches a contraption equipped with a series of tools: the sequence of hammer and chisel, pliers, pulleys, pylons and electric cables appears as a miniature assembly line. What is most surprising is not only the prefiguration of the processing of raw material directly in situ but how the architect is also the builder. He is a digital builder, thanks to the indispensable help of an electric (automatic) machine.

This first techno-enthusiastic imagination seems to have influenced the majority of automated construction research and development: many construction companies have developed production workshop as an actual automated construction factory capable of constructing full-scale building, even on-site[83]; many other have finalized systems that fabricate and assemble pre-manufactured components.[84]

Almost a century after the Villemard's vision—unusual for his time—the laboratories of the major architecture academic institutions are equipped with industrial robots that are seen as the medium increasing the control of the architect on the constructed object. By involving architects in the craft, the identity of the medieval master builder is integrated with a digital facet. In this advanced context of robotic production, the practitioners should be aware of the manufacturing equipment and they should design using the specific features of such machines.

---

[83] As pivotal example see the monster truck appeared on Mechanix Illustrated: the giant house-building machine is a remarkably forward-thinking idea for mass customization even though it doesn't involve any software yet. Reference to Houses While You Wait, (1946), *Mechanix Illustrated*, 6. Retrieved from http://blog.modernmechanix.com/houses-while-you-wait/#more [last accessed March 3rd, 2018].

[84] The ERNE Portalroboter (2015) is the most recent tailor made crane specially developed to built the Arch_Tec_Lab roof at ETH Zürich. For other bibliographic resource, refer to Scott Howe, A. (2000). Designing for Automated Construction. *Automation in Construction*, 9(3):259–276.

As a consequence, several issues are arising; first of all an issue of control, which is the characterizing feature of the master builder. The control includes both the ability to combine the different phases of the project and the know-how to solve designing issues; but also the control of the instruments. The technological development encourages architects to keep abreast of things and to look for integrative solutions that drive the project, not simply operatively but how the design becomes a process where everything is combined since the beginning. The digital master builder generates and provides instructions: to geometry, to material and to machines. The operative strategy collects and merges different intellectual resources in a serious experimental work and it is a full-fledged controlling process. Following the case in point of F. Brunelleschi, this new kind of master builder is thinking, designing and fabricating the technology, rather than simply using it. Control, furthermore, has also designing implications. As with the Villemard's architect,[85] what is important for digital master builders is not only the fabrication process but also the designing phase. In this sense, computational modeling appears as a controlling apparatus. Designers are establishing relationships, are assigning values to parameters, and are doing all the things that programmers do. It happens in a very different way from common sense about scripting, but nonetheless, it is programming.[86] What is crucial is to have the intellectual skills of abstraction and definition of relationship, but also the appropriate skills to manage parametric modelling software.

The idea of a digital master builder includes also the issue of authorship. In a sense, the architect is author[87] not only of the final output, but also of all the tools (both software, hardware and machines) involved in this combinatory process.

Finally, digital master builder requires three intellectual investments, or investigations: first, the venture in acquiring computational skills. The second is the investment in the design development for a particular CAD/CAM fabrication flow. And third is the occasional investment in material design. This last investigation is often the way to produce results that could not be achieved in any other way and is the pinnacle of the design-to-fabrication approach. The case studies are experimentation of these three attitudes under a light of a design experience at a smaller scale than a real building.

1.4.2.3 Digital Machines Involved in Architecture or Architectures Tailored for Robots

Introducing robotic manufacturing in architecture is revolutionizing the interface between computational design and physical materialization. Robotic techniques have

---

[85]The architect represented inside the control cabin has the plan, previously designed, in his hands. It means that the control of the process envisioned by Villemard it is not supervision but it starts since the designing phase.

[86]Program it is intended in its meaning of organize, and thus control.

[87]The Latin etymology *augere* means the ability of who makes it grow or makes it possible. Without the intellectual and manual involvement, the digital master builder could not manufacture his ideas.

been introduced in the architectural making only recently and they require designers to investigate the materialization process since the early design phase.[88] Given their character as programmable machines, which accurately perform movements,[89] robotic arm always need to be equipped with tailored end-effectors in order to perform high degrees of adaptability as architectural manufacturer.

There are two opposed but cogent cultural receptions in regard of the robotic fabrication capacity to innovate and make a difference in making architecture.[90] On the one hand, the inevitably fascination for new innovative technology produces enthusiastic fabrication innovations. On the other hand, a moderate assessment ensuing the technological improvements has also been developed. There are numerous historical examples of this kind of reactions; here an in-depth knowledge of tooling is needed in order to investigate the possible emergent failures.

The efforts of early pioneers in the field together with the adoption of open standards of robot programming have lowered the barriers to exploring the ingenious application of industrial robotics in architecture. However, the highly experimental character of applied academic researches and the predominance of traditional mass production techniques in building sites (SuperLab 2016) suspend the questions of immediate feasibility.[91] Instead, the concerted effort to meet the task of adapting industrial machines[92] to the architectural requests, and not vice versa, is the lively objective of digital crafting experimentations.

Aim of this sub-chapter is to introduce and describe robotic fabrication in architecture; the attitude is to warn against mistaken perspectives, which might lead into

---

[88] Industrial robotic arms are borrowed from the automotive and aerospace industry; their accuracy, flexibility, and reliability are enablers of computationally derived formal complexity.

[89] The accessibility of this technology to new users has also increased the request of algorithmic interfaces—often embedded in the same virtual design environment—in order to directly operate with the robot arm. It receives precise tool paths parameterized and scripted by architects. Common language communication occurs by plugin such as HAL Robotics ltd, KUKA|PRC, RAPCAM, Taco, Scorpion, rhino2krl but also own Python script are carried out. Retrieved from http://www.hal-robotics.com/ and http://www.robotsinarchitecture.org/kuka-prc and http://www.rapcam.eu/ and http://scorpion-robotics.com/ and http://blickfeld7.com/architecture/rhino/grasshopper/Taco/ and http://www.craftwise.ch/rhino2krl/ [last accessed January 17, 2018].

[90] Architecture schools, entrepreneurs, and artists have adopted robotic arm both for materialize computational architecture and for artistic performance e.g. Harvard Graduate School of Design (GSD) in Cambridge, Institute of Computational Design based in Stuttgart, Gramazio&Kohler Research at ETH in Zurich, Digital Fabrication Laboratory (DFL) in Porto, Carnegie Mellon School of Architectures in Pittsburgh, Hyperbody at Department of Architecture in Delft, REX|LAB Robotic Experimentation Laboratory at the University of Innsbruck, Fab-Union in Shanghai, ROBOCHOP in Munich or TOFA by the Berliner Chris Noelle, and Bot & Dolly based in San Francisco.

[91] However some pioneering actual-building-size realizations have already been accomplished. e.g. Gantenbein Vineyard Facade (2006), Landesgartenschau Exhibition Hall (2014), The Sequential Roof (2010–2016), to name only few.

[92] The reference is in particular to the well-known protagonist: the 6- or more axis multifunctional robotic plants.

mechanical fallacy[93] or automation impasse[94] that concentrate efforts solely in operational productivity (Gramazio and Kohler 2014). The true potential is to explicitly correlate the robot's generic nature both with architecture's specificity and material's production requirements, and not the other way around.

In this way, the robot reconsiders machining processes under an architectural light: the hands-on research with robotic arm questions the distinct and hierarchical separations between information and mechanics, technology and architectural culture (Gramazio et al. 2014b), or in other words, between speculative intentions and the engineering of robotic fabrication processes (Sheil 2012).

Since ancient time any mechanical or technological exploration and their resulting inclusion in everyday life's purposes has provoked dual reaction[95]: fascinated supporters, on the one side, confident of productive beneficial impact and detractors who caution about the margin between the thinkable and the feasible, on the other side. Architectural design routine usually starts with a given problem to be solved; the teamwork usually develops own tools to achieve suitable and unique results. This happens since ever: when skilled talents introduce economic, mechanic, or even social innovation they provoke reactions. Even Archytas[96] or the most popular Archimedes of Syracuse and Leonardo da Vinci[97] have caused opposing responses with their progressive discoveries or inventions. However, the route that architects are travelling in the last two decades is different. Pioneers have taken digital machines, as the industrial robot arm, and have started to explore its capabilities setting precedence and de facto initiating a new field of research in architecture. Behind a positive but sterile standpoint, marked by a belief that innovation has always-beneficial fallout, techno-utopianism[98] and enthusiastic stance[99] have undertaken multifaceted applied investigation (SuperLab 2016).

---

[93] Technical dedication might cause a fallacy in architectural achievements.

[94] It is referred to the aim of advancing futuristic facilities of making without referring to the architectural expression and a new aesthetic.

[95] "Accordingly, belief in progress and fear of progress would always confront one another in a certain context and in turn are mutually dependent; as would be expected, this oppositional pair forms an unbreakable and dynamic unit". Reference to Gramazio et al. (2014b), p 111.

[96] Archytas of Tarentum is a Greek mathematician reputed as the founder of mathematical mechanics. Retrieved from https://en.wikipedia.org/wiki/Archytas [last accessed January 17, 2018].

[97] Among all findings, it refers in particular to Codex Atlanticus where appears the drawings of what we can consider the first robotic design: a mechanism that features a front wheel drive, rack-and-pinion automobile with the ability to control its own motion and direction.

[98] This topic is also well argued by Scott F D (2010) Architecture or Techno-utopia Politics After Modernism, MIT Press, Cambridge.

[99] The Economist's front cover previously mentioned clearly alludes to an enthusiastic and fascinated belief for 3D printing as a revolutionizing technology. Reference to Print Me a Stradivarius. (2011, February 12). The Economist, 398(8720). Retrieved from http://www.economist.com/node/181143 27 [last accessed January 17, 2018]. Going back in time, the historical revolution of printing during the Renaissance and its revolutionizing effect on society, or also the more architectural related revolution foreseen by R. B. Fuller comes immediately to mind as prime examples of this attitude. His endorsement of technological progress that he saw as a means to redesign society find an accomplishment in the on-going widespread research line in robotic materialization of architecture. Furthermore projects, conceptual work or publications emerged in the late 1960 s have introduced

The industrialized chain of production and automation (Andia and Spiegelhalter 2015) are pervaded by utopian concerns.[100] At first sight, the advent of robotics in architecture remarks the emergence of similar respects already brought up by former Modernist attempts. The employment of industrial fabrication facilities solely for the purpose of generating complex geometries and renderings warrants some scepticism. Techno-pessimists warn against the possible transformation of architectural research into a fully automated and rationalized production industry (SuperLab 2016). All of this aside, it is easy to measure the operative limitations of robot on ordinary building site. Innovative construction techniques present their own difficulties, as does 1:1 scale fabrication using robot. Besides the incompatibilities with traditional construction sites, security and maintenance are also determinants because robotic arms are not developed for working outside their controlled environment.

The quest for a new seamless continuum grounded within computation between the architect's mind and the built reality is without any other precedent. This is the turning point. Industrial machines used to perform a customized materialization and to assemble parts have been interpreted as tools radically distinct from the mind that make them work. "What is under way is a comprehensive development" (Gramazio et al. 2014b) of robotic fabrication, which directly confronts architects with the need to cooperate with the technological equipment, rather than solely being fascinated users. The simplistic use of advanced tools may result in the attribution to what robotic labs succeed to manufacture of a contrived architectural label.

In reversing the design routine, pioneering architect's attitude goes beyond the limits of what the modernist industrialization has already dreamt and achieved: that is, force the architectural exploration into the boundaries of a rigorous streamlining. In fact, the automated fabrication processes of industrial robot are activating innovatory features, enabling the fusion of computational design and constructive materialization as the hallmark of architecture in the post-digital age.

References

Alexander C (1964) Notes on the synthesis of form. Harvard University Press, Cambridge
Anderson C (2012) Makers the new industrial revolution. Crown Business, New York
Andia A, Spiegelhalter T (eds) (2015) Post-parametric automation in design and construction. Artech House, Boston
Barazzetta G (ed) (2015) Digital takes command design horizons and digital fabrication. Rubbettino, Catanzaro
Booth P (2009) Digital materiality: emergent computational fabrication. In: 43rd Annual conference of the architectural science association, Tasmania

---

the remarkable ambition that architecture could be ethically relevant. Reference to Scott FD (2010). Architecture or Techno-utopia: Politics After Modernism, MIT Press, Cambridge, and SuperLab (2016).

[100]Such as the desire to find a match between nature and technology or the idea to free man of unnecessarily harsh handwork.

Brell-Çokcan S, Braumann J (eds) (2013) Robotic fabrication in architecture, art and design 2012. Springer Wien, New York

Burry M, Burry J (2016) Prototyping for architects. Thames & Hudson, London

Caneparo L (2014) Digital fabrication in architecture engineering and construction. Springer, Dordrecht

Computer Software (1984) The magic inside the machine. Time 123(16)

Corser R (ed) (2012) Fabricating architecture selected readings in digital design and manufacturing. Princeton Architectural Press, New York

Eisenman P, Gehry F, Gianni B, Hoberman C, Lynn G, Yoh S (2013) Archaeology of the digital. Sternberg Press, Berlin

Fok WW, Picon A (eds) (2016) Digital property open-source architecture. Wiley, London

Frampton K (1995) Studies in tectonic culture the poetics of construction in nineteenth and twentieth century architecture. MIT Press, Cambridge

Frazer J (1995) An evolutionary architecture. Architectural Association, London

Frazer J (1999) Towards the post digital era. In: AVOCAAD second international conference proceedings, Hogeschool voor Wetenschap en Kunst, Brussels, pp 35–39

Gershenfeld N (2012) How to make almost anything the digital fabrication revolution. Foreign Aff 91(6)

Gramazio F, Kohler M (eds) (2014) Made by robots challenging architecture at a larger scale. Wiley, London

Gramazio F, Kohler M, Hanak M (eds) (2008) Digital materiality in architecture. Gramazio & Kohler, Müller, Baden

Gramazio F, Kohler M, Willmann J (2014a) Authoring robotic processes. Architect Des 84(3):14–21

Gramazio F, Kohler M, Willmann J (eds) (2014b) The robotic touch how robots change architecture. Park Books, Zürich

Grobman Y, Yezioro A., Capeluto I (2010) Non-Linear architectural design process. International Journal of Architectural Computing 8(1):41–54. https://doi.org/10.1260/1478-0771.8.1.41

Jordan J (2016) Robots. MIT Press, Cambridge

Kanaani M, Kopec DA (eds) (2016) The Routledge companion for architecture design and practice: established and emerging trends. Routledge, New York

Kelly K (2011) What technology wants. Penguin books, New York

Kieran S, Timberlake J (2004) Refabricating architecture how manufacturing methodologies are poised to transform building construction. McGraw-Hill, New York

Kolarevic B (2003) Architecture in the digital age design and manufacturing. Spon Press, New York

Kwinter S (2003) The computational fallacy.Thresholds 26:90–92

Kwinter S (2007) Far from equilibrium: essays on technology and design culture. Actar-D, Barcelona, New York

Lorenzo-Eiroa P, Sprecher A (eds) (2013) Architecture in formation on the nature of information in digital architecture. Routledge, Abingdon

Menges A (2008) Integral formation and materialisation: computational form and material gestalt. In: Kolarevic B, Klinger KR (eds) Manufacturing material effects rethinking design and making in architecture. Routledge, New York, pp 195–210

Menges A (2009) Uncomplicated complexity integration of material form, structure and performance in computational design. In: Hirschberg U, Bechthold M, Bettum J, Bonwetsch T, Bosia D, Cache B, Vrachliotis G (eds) Nonstandard structures. Springer, New York, pp 140–151

Menges A (ed) (2012) Material computation higher integration in morphogenetic design. Wiley, Chichester

Menges A, Ahlquist S (2011) Computational design thinking. Wiley, Chichester

Naboni R, Paoletti I (2015) Advanced customization in architectural design and construction. Springer International Publishing, Cham

Negroponte N (1995) Being digital. Knopf New York

Pask G (1969) The architectural relevance of cybernetics. Architect Des 39(6/7):494–496

Picon A (2010) Digital culture in architecture an introduction for the design professions. Birkhäuser GmbH, Bâle

Pine BJ II (1992) Mass customization the new frontier in business competition. Harvard Business School Press, Boston

Riiber J (2011) Generative processes in architectural design. The Royal Danish Academy of Fine Arts Schools of Architecture, Design and Conservation, Copenhagen

Sheil B (2005) Transgression from drawing to making. Arq Archit Res Q 9(1):20. https://doi.org/10.1017/s1359135505000059

Sheil B (ed) (2012) Manufacturing the bespoke making and prototyping architecture. Wiley, Chichester

SuperLab (2016) The nature of robots robotic fabrication in architecture, CreateSpace Independent Publishing Platform

Sutherland IE (1963) Sketchpad a man-machine graphical communication system. PhD thesis, Massachusetts Institute of Technology, Cambridge. http://hdl.handle.net/1721.1/14979

Tedeschi A, Wirz F (2014) AAD_Algorithms-aided design parametric strategies using grasshopper. Le Penseur publisher, Brienza

Terzidis K (2006) Algorithmic architecture. Architectural Press, Oxford

Thomas KL (2007) Material matters: architecture and material practice. Routledge, London

Thompson DW (1917) On growth and form. Cambridge University Press, Cambridge

Weiser M (1991) The computer for the 21st Century. Sci Am 265(3):94–104

Chapter 2
Material: Digital in Action

2.1 Experimental Opportunities

Architectural design is occasionally meant as a symbolic dialogue, rather a questioning, between project's requests and respective solutions. The route to a suitable design winds through attempts, incomplete representations, or models: the experience of travelling back and forth enriches each choice and architects seek feedback from them in order to produce working prototypes. Developing and building different versions of one single design or simply parts of it represent an opportunity that helps the final choice about the kind of material, detailing of joints or principals of assembly. This kind of approach is a common thread combining all the experiences presented in this chapter.

Making models and 1:1 scale prototypes is a tool for thinking in itself for scholars, who reconsider digital fabrication technologies as an interface between the digital and the physical world. And there's more. As this little catalogue conveys, the customization of computational design and manufacturing tools has controverted the vision of architectural design from a preemptive act into a preeminent experimental process.

> [...] all models are wrong,
> but some are useful.
>
> G. E. P. Box, 1976

If we extrapolate from the context of statistics the aphorism by G. Box and attribute to the model its customary architectural meaning as the material manufacture of a representation—instead of an ideal and abstract construction intended for inference—we certainly appreciate the suggestion of the British statistician about the usefulness of models, so intended. The long-lasting activities surrounding design and manufacturing have historically combined tools, models and materials. The parallel advancements in each of these fields have increased the possibilities in strengthening the bond—occasionally the boundary—between the drawn and the made. In this

© The Author(s) 2019
A. Quartara and D. Stanojevic, *Computational and Manufacturing Strategies*,
SpringerBriefs in Architectural Design and Technology,
https://doi.org/10.1007/978-981-10-8830-8_2

light, the human act of building arises as a final fulfilment from the mutual dialogue between the designing idea and the model.

Historically, especially before the advent of computation, the value of a project has been transmitted through its model. The renowned episode about the Florence Cathedral's dome award of contract, dated back in March 1420, epitomizes the virtue of prototype in architecture. The young goldsmith F. Brunelleschi, nominated master builder as consequence of the public competition, presented his innovative constructive ideas through wooden models. This perfect example is the evidence of how the material model brings with it a considerable value, above all for the constructive process and not only for the aesthetics of architecture.

During the early 1960s, the Massachusetts Institute of Technology has pioneered this practice,[1] initiating the redefinition of the historic relationship protocols between design and construction and leading the advent of the designer-as-maker's trend. Slowly and progressively, the architectural chain leaves behind the sequential protocol that splits thinking and doing.

The experimental exploration of innovative computational manufacturing approaches is crucial in combining design process and material consideration with their production. The bespoke becomes the distinctive character of both material outcome and fabrication hardware (Sheil 2012). On the one hand, the materialization of complex geometries expresses the iconic, aesthetic and innovative uniqueness by its physical appearance. While on the other hand, fabrication tools' development makes own novel demands. The involvement of CNC fabrication technologies within research experiences developed by schools of architecture claims proper customizations, until tools' design and their fabrication becomes a project itself.

The variety of the off-the-shelf machineries has been considerably expanded in the last decades: some material-specific tools as well as innovative manufacturing techniques are remarkably increasing the possibilities of digital fabrication facilities. By necessity, the training of young architect at the technical level of machineries is indispensable: "As architects we need to engage directly with machines, from computers to instruments of physical manipulation". In order to operate machining processes, a domain of communication language is needed: to all effects, it is the means through which the seamless computational flow reaches its physical actuation. Sets of customized manufacturing tools (i.e. end effectors) and proper instructions for machines (i.e. G-code[2] and robotic motion path) are determinant constituents of this new paradigmatic way of making in architecture. Computational talented architects are looking beyond ordinary fabrication technologies. The intimate knowledge of servo control, communication protocols and tooling parts results in the fabrication of custom tools. As M. Burry suggests, the digital fabrication advent has taken the

---

[1] First CNC machines were developed as prototype by MIT and the first CNC lathe was introduced to the market in 1952. During the 1960s, CNC machines start to circulate aviation, shipbuilding and automotive industries. Only in the last two decades, CNC machines and robot significantly have spread out in architecture's academia.

[2] G-code is a language in which people tell CNC machine tools how to make something. Generally, there is one international standard: ISO 6983.

credit of blurring the model and prototype boundaries (Sheil 2012), focusing on a discipline that intertwines design and production.

The case studies experiment this methodological renewal through wooden structures. Wood appears as one material, but its numerous varieties give rise to manifold products and alternatives to experiment with. Wood is the porous and fibrous structural tissue of trees thus it is a heterogeneous and hygroscopic material; its chemical composition varies from species to species but it has four main components: water, cellulose, hemicellulose and lignin. It is made up of cellulose fibres, which have high ability to resist compression forces. The structure of both hardwoods and softwoods, the two categories in which all the wood species are grouped, differs considerably in regards to the cellular structures. The cells' distribution determines most of the resulting properties, characteristics and behaviour of wood. As consequence of the specific structure, distribution and orientation of the fibre cells, wood is an anisotropic material. Instead of conceiving this property as problematic, woodworkers have developed several strategies to reconcile the biological irregularities with constructive requests.

Moreover wood demonstrates important advantages over most commercial semi-finished construction materials: for example, it has unsurpassed ecological virtues that enlarge its application's interest. Wood has been a vital production material since humans began building shelters, utensils and means of transport. Usually, the wooden elements intended for the construction and the most varied support applications take the specific name of timber. Nowadays, these wood products, historically and widely used in construction industry, restate the influential role that timber has during material-constrained design, especially combining structural and decorative expressions within a computational path (Beorkrem 2013).

Architects are honing robotic and CNC fabrication techniques for manufacturing purposes reconsidering wood and its derivatives as suitable building materials for the experimental design researches at an actual scale and no longer as an out-of-date material. Recent years have seen unprecedented innovation of new technologies for "advancing wood architecture" (Menges et al. 2017). Frequently, timber elements are regularized components such as sticks, blocks or panels: these off-the-shelf wood products come in manageable dimensions, but also tree forks conform to different digital processing techniques, ranging from rough chainsaw cuts to detailed milling carving or sophisticated chemical modifications (Beorkrem 2013).

The manifold availability in both wood products and natural grain woods is tested and exploited in each project; these latter provide evidence of the outstanding role of wood in relation to computation and digital fabrication. Typically, wood products have a vast set of capacities allowing both industrial design and architectural scale detailing and assembly. Common topics of works are connection optimization, structural properties tests, as well as environmental performances in a broad sense (involving resource management, cost-effectiveness and behaviour reactive to environmental stimuli). Usually wooden elements come from the stems of trees. But the innovative use of 3D printers makes possible the reuse of production discard or sawdust as finishing material, enriching the environmental sustainability values of wood. The digital turn in timber manufacturing during the last decade has certainly offered new possibilities, increased precision and has allowed the non-standard test and construction of innovative wooden structures.

Material culture has relevance for studies of the past, but it also brings a projective capacity within the computational workflow, breaking the conceptual separation between the processes of design and the physical making of the built environment. In stark contrast to previous linear and mechanistic modes of digital manufacturing, the digital materialization is now starting to coexist with design as an explorative "cyber-physical" (Menges 2015) process. Fabrication-aware design is reawakened not in the sense of the tenet "truth to materials"[3] of modernist attempts, but rather as a truly generative material-to-fabrication exploration.

Encouraged by the insightful activity of precursors,[4] such as A. Gaudí or F. Otto, the material computation duo becomes an emerging research field during the twenty-first century. Those studies have laid the basis for what is defined an "extensible framework", which does not linearly set procedures, but acts as an "interoperable design event" bridging virtual and real (Ahlquist and Menges 2015). The ever-accelerating convergence of the computational and the physical radically transform researches, which directly aim to make tangible the bit.

The computational practice of bringing the digital into action is matter of designing possibilities. Despite its deep involvement with logic, it does not determine local and specific interaction. Rather, it is established as a variable framework, which allows a range of specificities between materiality and form. This opportunity for architects occurs when the traditional design and fabrication processes are rethought[5] (Thomas 2007).

Over the past decade, several researches have paved the way for future development in building technology; in particular academic institutions have extended the computational design flow to its materialization within elective courses or masters. Academic researchers as well as open-minded companies are pioneering this trend connecting digital outcome with robotic materialization. The numerous and highly

---

[3] This often-quoted statement relates the assumed essence of a material to a set of given structural and spatial typologies. It holds that any material should be used where it is supposedly most appropriate and its nature should not be hidden, thus it should be left unfinished and exposed. This belief was closely associated with the Arts and Crafts Movement.

[4] Three predigital parametric personalities are taken as case studies. A. Gaudí and his *magnum opus*, the Basílica i Temple Expiatori de la Sagrada Família in Barcelona, J. Albers with his Vorkurs course at the Bauhaus in Dessau and F. Otto and his extensive series of material experiments in Stuttgart. At their time, they provided excellent examples where the projective capacity of material was explored by scale models. In different ways, they all evidently reveal the exploratory horizons embedded within the materially informed design. The experimental approaches were undertaken not as symbolic and scalar representations of engineering constructs, but employed material as a driving force for developing new open-ended architectural designs, rather than optimized forms. Reference for further details on form-finding are: Horowitz F, Danilowitz B (2009) *Josef Albers: To Open Eyes*, Phaidon, New York; Otto F, Rash B (1996) *Finding Form: Towards an Architecture of the Minimal*, Edition Axel Menges, Stuttgart.

[5] "The structural production process that emerges in digital materiality is no longer that of the construction site or the workshop but rather a design process according to specific guidelines of the architect" Reference to Willmann et al. (2013).

experimental works exploring this deep reciprocity often defer the resistance to innovation of the building construction sector and they develop a decidedly innovative way to fabricate.

On the one hand the "digital by material" aptitude is permeating in today's discipline as a "not incidental" and an "interdependent structuring" that can be "analytically developed and implemented on an architectural scale". On the other hand, it is producing "enticing manifestations" (Willmann et al. 2013), which do not insist on fabrication but rather on ornamental purposes.[6] By the increasing integration of material and production as two driving agents within the computational procedures, and thanks to the wide spreading of digital fabrication technologies, new trajectories for the construction industry as well as for designers are emerging. Beside the experimental nature of these projects, the pioneering role of the research arises: it aims to push the boundaries towards industry and it is questioning how far these researches are from releasing frameworks that can be used in building construction field.

The works of post-digital architects are expounded as the synthetic merging of interdependent values; under this light, the case studies aim to demonstrate how to avoid the self-referential and excessively abstract computational workflows. It embarks on a showcase of academic activities, proposing tools and frameworks for researchers to evaluate tendencies. Definitely, the projects involve both material logic as well as the logistics of materialization: the latter constitutes the real novelty of the digital in action since it simultaneously manages an unusually high number of elements and automation.[7] While documenting research projects, this chapter carries a critical standpoint, drawing attention to certain limitations that represent the fields of future developments. Case studies are gathered in two main groups, collecting them according to material tolerances and material enhancements. Each section introduces some hypotheses and concepts that the projects explore in an experimental way.

2.2 Material Tolerances

Before addressing the details of the case studies gathered within this paragraph, introducing the concept and the meaning we attribute to material tolerances is primarily needed. First of all, they differ between the digital and the physical worlds; the first focuses on the design of objects perfectly matching for assembly with no gaps, while in the physical the friction of the material defines the additional machining of edges to allow the pieces to come together. And then tolerances. As the dictionary suggests us the word tolerance acquires different meanings according to the context.

> tol·e·rance/ˈtɒlərəns/noun [...] 2) capacity to endure something, esp. pain or hardship. 3) the permitted variation in some measurement or other characteristic of an object or workpiece
>
> Collins English Dictionary

---

[6]Sometimes applications appear as artistic supplement rather than material studies.
[7]These are the distinguish characters of mass customization. See *Mass-Customized Material Accomplishment*.

For the purpose of our debate, we have to transfer the usual linguistic meanings to a set of computational values and features,[8] and definitely custom characterizations that tolerance may take on in our case studies. The great variety of linguistic definitions takes on various facets in the different experimentations; in relation to whether project phase, materials, machining processes and assembly logics tolerance assume three different meanings.

From a digital drafting and modelling point of view, tolerance management often is solved with units, objects snap and precision settings that every CAD software has on its properties tabs. A careful modeller can virtually manage different conditions (i.e. perpendicular, adjacency, intersection, projection, tangency, centring and so on) of vertices, edges or planar and curved faces. On the other hand, from a constructive point of view, every detail of architectural fabrication depends on the control of the same dimensions and constraints. Their management is not as simple as it is within a virtual environment through switching on and off options on parametric design softwares. It is not the case of an indecipherable possibility that certain software make possible as suggested by the aforementioned title of the magazine Time. [9] In this first sense–that is also the most common and direct—tolerance is strictly connected to sizes, distances and any kind of measurements related to pieces of material, objects or movements.

Another aspect of this meaning relating tolerance with measurements is tolerance as safety. We ascribe to tolerate the ability of the entire production chain to adjust robot movements, cutting parameters, minimum distances in order to avoid collisions, waste or inaccuracies.

One last acceptation describes tolerance as a permitted variation that allows little but substantial changes to a predetermined condition. In this sense, the researchers who developed the case studies while acting the part of digital master builders[10] have generated tolerance as the ability of the code in choosing and adapting some operations. The different algorithmic codes allow, within narrow and often well-defined boundaries, a deviation from what was previously set. They allow parameters correction on local level, which directly affects the global system, updating instantly; it brings the possibility to move in a loop from local to global withing a synchoronized system. Ultimately, this meaning of material tolerances concerns all the cases when it allows or admits that something is done, or happens.

---

[8]It is important for the reader to remember that the dissertation at the start of this book is about features and their possible connotations.

[9]Reference to Computer Software The Magic Inside the Machine (1984) *Time*, 123(16).

[10]Reference to the previous paragraph *The Digital Master Builder*.

2.2.1 Design Chain for Discrete Engineered Wood Elements

2.2.1.1 Previous Experiences

During the first decade of the 2000s, academic institutions have started to carry out several mid- and long-term courses engaging robotic fabrication as an active design parameter. On the educational level, tutors and students have been majorly exploring the bespoke design path for wooden applications. It's to say the wood field has been the most explored. Building components, temporary structures and full-scale pavilions started to populate the academic scenario; Gramazio Kohler Research Group (ETH, Zurich), Institute "for Computational Design (ICD, Stuttgart) and IBOIS (EPFL, Lausanne) are the leading schools that pioneered the studies on this field.

Four particular projects developed in Zurich are elected as references since they used engineered wood elements—mainly sticks and slats—of small size. The Stacked Pavilion (2009), The Sequential Structure 2 (2010), Shifted Frames 2 (2013) and Complex Timber Structures 2 (2013) are investigations that materialize complex geometric configurations combining design prerequisites with their CNC manipulation.

Starting from the analysis of contemporary and traditional timber construction systems, the different research teams from ETH have driven forward several projects liberating "novel aesthetic and a functional potential" (Menges et al. 2017). Whitin this context, they have developed robotic fabrication processes opening a completly new path to architecture, for istance, they have implemented complex geometrical assembly of small-scale wooden beams, and multidirectional stucking of wooden elements originating structurally optimised shelters. Definitely, the fully automated assembly operations establish a new expression on the traditional wood material. These robotic construction experiments are assembly-driven processes; sometimes are integrated with manual assembly and they are adopting industrial logic of assembly automation.

The case studies presented in this paragraph (namely Fusta Robòtica, Digital Urban Orchard and the Foundry building) are aimed to present the goal to not transform the constructive process into a fully automated procedure, but into an interlacement of the strengths inherited from craftsmanship and from robotics, combined as designed parameters even at an early stage.

2.2.1.2 Robotic Layer-Based Assembly Systems

The present-day investigation occurring within the digital architectural scenario drives several researches through complex and fascinating geometric generations. The following case studies demonstrate how remarkable small-size spaces can be characterized by differentiations through serial variation combining material constraints with simple geometric rules. The following case studies are experimentations—part of the same Applied Research program called Open Thesis Fabrication

Fig. 2.1 Digital and virtual visualization of the Fusta Robòtica Pavilion

2015 (OTF 2015) developed at the Institute for advanced architecture of Catalonia (IAAC, Barcelona)—testing how architectural objects gain coherence from the local differentiation of their components. In particular, Fusta Robòtica and Digital Urban Orchard employ different square sections and straight wooden sticks in countless different lengths of two essences exploiting their geometric shape and expanding their restraints.

Fusta Robòtica[11]—meaning in Catalan robotic wood—is developed in occasion of the Setmana de la fusta de Catalunya 2015 (Fig. 2.1), the most important event in the world of wood and furniture in the Catalan territory and it was used as a testing ground for familiarizing with available technologies of the Iaac workshop. This case study is a challenge turning the limits of Catalan wood[12] in possibilities through the application of parametric design and robotic fabrication processes. The collaboration with Serradora Boix (Fig. 2.2), the local company that supplied the wood, brought to the attention of the project the potentiality of a quasi-zero-kilometre material obtained through sustainable forest management. The company has developed its own chain for the production of primary processing of wood (lumber, chip, bark and sawdust) and offers the opportunity to build a stable, large-scale prototype by using the low-impact Catalan wood.

---

[11]The project has been conducted during a short term by an international team of students and researchers. Credits go to Iaac and the team composed by faculty: A. Markopoulou, S. Brandi, A. Dubor, D. Stanojevic, and students: J. Alcover, A. Figliola, Y. Haddad, J. Won Jun, M. Kumar, M. M. Najafi, A. Quartara, F. S. Shakir and N. Shalaby. The pavilion would not have been possible without the generous sponsorship of Serradora Boix, in collaboration with, Gremi de Fusters, Tallfusta, Incafust, Mecakim and Decustik.

[12]Local wood unfits structural applications because of the often-irregular fibre arrangement and it is solely used for palettes, i.e. wood knots.

Fig. 2.2 Wood storage at Serradora Boix

The pavilion is the actual materialization of a one-sheet hyperboloid, also called hyperbolic hyperboloid. It is a particular double-curved surface which has a negative Gaussian curvature at every point and, more importantly, it is a ruled surface. The mutual dialogue between geometry and the available material encouraged the researchers to exploit, among its properties, the possibility of drawing straight lines on the hyperbolic hyperboloid. Furthermore, the opportunity to obtain a circular section while intersecting the surface with vertical planes allowed a discretization[13] process achievable through wooden sticks. Taking into consideration this geometric analysis, the scholars pondered the parametric generation of a digital model in order to evaluate and arrange the fabrication and the assembly logics. Instead of following the common surface generation obtained by rotating a hyperbola around one of its principal axes, the parametric modelling started from a cylinder (Fig. 2.3a). In fact, beginning from a cylinder with its main axis lying horizontally, the parametric definition gradually rotates one of the two base circles (also named directrix) up to 80° (Fig. 2.3b). The two base circles of the cilynder can be subdivided with a series of points; these points can be then connectted from one circle to the other with straight lines. When one of the two circle rotates, the straight lines follow the rotation changing the morphology of the cylinder to a hyperbolid hyperboloid. The process is carried out in parallel with an inner offset surface. Once the final shape achieved, a

---

[13]In mathematics, discretization is the process of transferring continuous functions, and equations into discrete counterparts. In this case, the continuous and connected one-sheet hyperboloid surface is approximated through the wooden sticks.

vertical plane intersects the sets of generatrix (Fig. 2.3c) and a pattern of cross intersecting segments is generated. The process is then restricted to the solely half above
the ground plane (Fig. 2.3d). The two-dimensional elements become B-reps[14] virtual
representation of the actual wooden stick with discrete height, length and depth. The
scissor-like pattern (Fig. 2.4) is used to discretize the one-sheet hyperboloid and the
number of sticks along the circle directrix is set according to cutting length tolerable
by the robotic fabrication set-up. The series of alternated triangulations and the use
of green wood have been safety precautions used to construct a redundant structure,
capable of absorbing the deformations produced when drying. So far in the project
description has been explained the geometric representation, but it still misses the
explanation of the integration in the design process of material characteristics and
the robot workspace. The robotic assembly of unprocessed wooden sticks, whose
orientation and density can be controlled with the aid of the parametric model, is in
the foreground of this short-term experimental project.

All the design-to-fabrication phases of the Fusta Robòtica Pavilion had to deal
with calibration and tolerance settings that would meet the wood irregularities. The
primary wood processing carried out at the Serradora Boix sawmill produced some
inaccuracies in the stick section. Sticks generally had a square section of 38×38 mm,
but in reality, many of these were slightly bent or had irregular rectangular sections
that distanced themselves from hypothetical measures. Thus, the movements of the
robot and end effector have been studied to consider such tolerances and compensate
them where possible. Manufacturing logistics was also considered as design tolerance; considering the dimensions of the work platform of the robot cell, the pavilion
was designed to be built in 8 arches, each divided into two halves. In total, the 16
components were assembled and nailed in a horizontal position following the layer-
based logic (Fig. 2.5). The fabrication lasted 35 h and combined the robotic pick,
cut and place loop with manual nailing. At the end, the pavilion is made up of 940
different sticks; it spans almost 4 m from front to back and around 5 m side to side
and reaches a height of about 3 m at its highest point. The prototype is now located
at the Valldaura Labs in the Collserola Park (Fig. 2.6).

The know-how developed during this experience has driven the scholars in designing the Digital Urban Orchard (Fig. 2.7) as a more in-depth and improved research. It
is a wider designing experience rather than a robotic fabrication exercise; it includes
concepts as digital mass customization,[15] a new concept of socialization space and
food production. The new kind of urban agriculture is based on hydroponic and
it is strictly related to the shape optimization, generated according to environmental parameters and other apparatuses, as the silicone skin. The pavilion is meant to
be the first sample of a series of digital urban orchards populating the rooftops of
Barcelona. By adapting the footprint and optimizing the surface from time to time,
each greenhouse might take advantage of the custom design workflow in conforming

---

[14]In solid modelling, a boundary representation (B-rep) is a model composed by topology (faces,
edges and vertices) and geometry (surfaces, curves and points).

[15]It can be defined as the use of digital tools to mass-produce variations at no extra costs. A more
in-depth definition can be retrieved in Sect. 1.3.2.2 *Mass-Customized Material Accomplishment*.

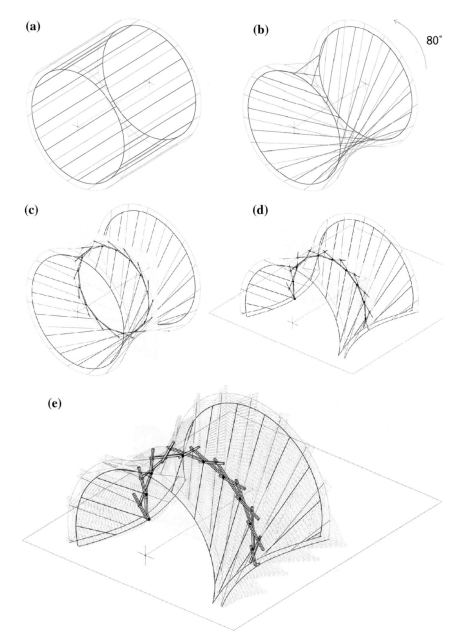

(a)

(b)

80°

(c)

(d)

(e)

Fig. 2.3 Generation of form: cylinder (**a**), rotation (**b**), vertical intersection (**c**), horizontal trimming (**d**), final discretization (**e**)

to the available space on any particular flat rooftop. Therefore, it was necessary to generate a redundant construction, which allows a very high degree of adaptability

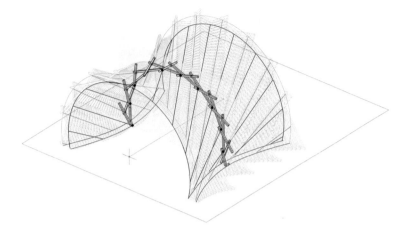

Fig. 2.4 Scissor-like pattern discretizing the shape through repetitive triangulations

Fig. 2.5 Robotic fabrication: wooden components ready to be assembled

but at the same time evaluates aspects of saving materials in the phase of machination and repeatability.

The simultaneous concerns on designing a stable structure and on gaining the solar radiation have required multiple design expedients in complying with each one of the functional, structural and environmental performance criteria needed within a greenhouse. A series of environmental analysis has made available a collection of data related to the site location. The parameters that have been taken in to account

Fig. 2.6 Fusta Robótica Pavilion reassembled at Valldaura Labs

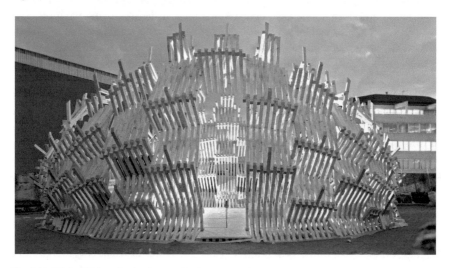

Fig. 2.7 Digital Urban Orchard assembled on the rooftop

were mainly the solar radiation and the wind (average speed and direction); they were analysed during winter and summer, considered as the most critical periods of the year for growing plants. In particular, the qualitative analysis on prevailing wind average

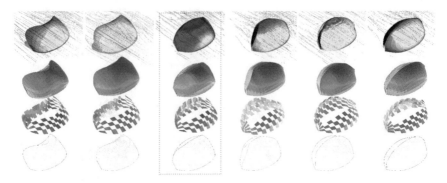

Fig. 2.8 Catalogue of six different shape alternatives

velocity and direction was considered at the level of the actual construction site (the rooftop is located about 25 m of height). The incident radiation value extracted from the solar analysis has guided the shape generation and its optimization. A catalogue of different buildable surfaces (Fig. 2.8) has been generated through an evolutionary algorithm. Through the adaptable scissor-like pattern, the stick assembly alternately offers flat supporting areas for the hydroponic pipes, either it constitutes space-functional furniture or some extended sticks are designed as holder of the silicone skin. The density of the structural pattern also responds to optimization logics for solar gain and considers almost total transparency at the top of the pavilion. The algorithm was able to calculate the global shape, deforming within a given range two basis curves positioned on the ground, and a matching curve (Fig. 2.9a). At the same time it was reorienting the plot areas, maximizing their size and optimizing their positioning. The result of this optimization process was then analysed in terms of the wind pressure on the surface (Fig. 2.9b) in order to verify the global stability. This cocoon-shaped design (like a coccon) choice has a footprint of about 27 m^2 (floor covered area), it is 8 m long and almost 5 m "wide", with a planting area of about 20 m^2 (the planting area is the space dedicated to growth of vegetables) and a covering surface of about 75 m^2 (the covering surface is the outer skin). "The surface has been discretized with a pantograph-like module" referes to the scissors-like structure, the triangulated arch is then traferred geometrically to a pantograph-like connection (Fig. 2.10). The cocoon-shape hides in between a misleading undifferentiated amount of Redwood sticks manifold structural purposes and functional ones. The square section sticks (45 × 45 mm) are distinguished into different functions: main trusses (Fig. 2.9c), plant supports, skin holders (Fig. 2.9d) and furniture (Figs. 2.9e) and (2.10).

The decision to purchase wood at the Gabarró company and to not use anymore the zero-kilometre material has made it possible to increase manufacturing precision. The solid wood strips of Scots pine—known as Flanders in certain areas of Spain—have been provided with four-sided sharp edges. Even though the supplier's technical sheet points out frequent defects (e.g. small to large knots or small resin bags), the

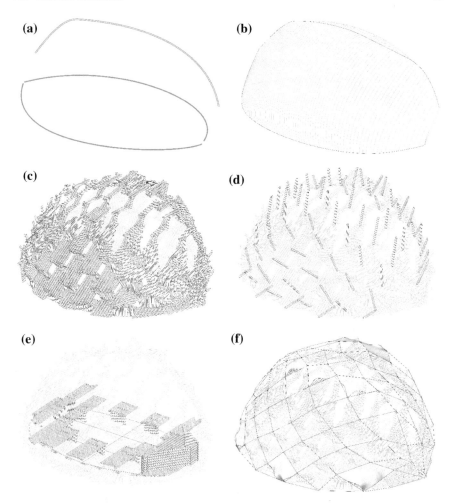

Fig. 2.9 Shape generation: starting horizontal and vertical curves (**a**), final surface (**b**), sticks used for main trusses (**c**), sticks used as skin holders (**d**), furniture components (**e**), an overview of the completed pavilion (**f**)

cutting process in bars is much more accurate and all the sticks can be considered with a constant section in their length and also the same between different sticks. After several calibration tests, it was possible to achieve a precision of the robotic processing (picking, cutting and positioning) with tolerances lower than one-tenth of a millimetre (Fig. 2.11). The wood was provided in 4200-mm-long bars; according to the average length of the final sticks, it was needed a manual pre-cut, also because the robot combined with its end effector cannot manage such lengths within its room. Three cutting patterns have been designed in order to build a custom wood dispenser

N   S

Fig. 2.10 Section showing the interior functional arrangement

accommodating sticks of 1500, 1000 and 600 mm in length and reducing the waste of unprocessed material.

Having gathered these steps in the structure of a parametric code, the team developed further the algorithm generating the G-code for the production phase including three additional dimensional tolerances whose calculation was totally automated for each one of the 1680 sticks. Each wooden stick is handled throughout the robotic fabrication loop: picking, cutting and placing (Fig. 2.12). According to the respective final sticks positions and functions, they are selected from one out of the three starting sticks, provided by the custom-made wood feeder. The picking position is automatically adjusted according to the safety distances between the saw's blade and the metal part of the gripper (Fig. 2.13). They are then cut in various and always different lengths and their end edges are shaped with different three-dimensional angled cuts. The circular saw (it can be equipped with blades up to Ø 260 mm) is fixed onto the rotary table allows β and γ angles while cutting (Fig. 2.14). Furthermore, the positioning has been calibrated as an interlocking movement avoiding possible collision with stick already positioned but not nailed (Fig. 2.15). Dimensional tolerances have been crucial and took several calibration tests (Fig. 2.16). Thus each stick, varying each time the fabrication loop, informs the robotic fabrication code. After a layer is processed, each stick is manually nailed with the help of an industrial pneumatic nail gun[16] (Fig. 2.17).

According to robotic fabrication constraints and manual assembly logic, the final symmetric shape has been split into 12 components, 6 types of sections on the left side, and the same ones but mirrored on the right side (Fig. 2.18). Three manufacturing strategies have been defined depending on the size of the sections. They involve the

---

[16]Shank spiral nails were chosen in order to increase friction with wood.

(a)   **(b)**

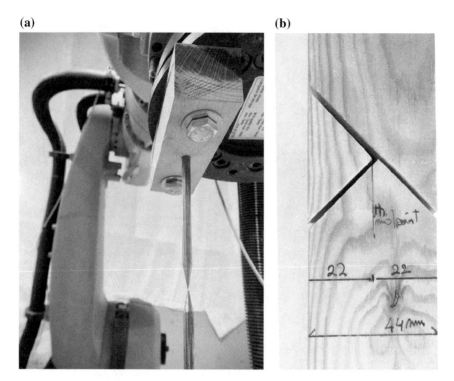

Fig. 2.11 Robot calibration (**a**); test cuts performed to verify the precision of the cutting set-up (**b**)

robotic processing of the entire section or of two halves or of three parts of the final arc section with 30 assembled parts in total. After the 52 h of production within the robotic room, the sections were lifted with a jib crane on the rooftop and assembled during 36 h of unskilled manual assembly (Fig. 2.19). This custom-made digital design work flow gives rise to a rapid and automated production process, with only 2% of scraps from the material supply.

Regarding the covering of the greenhouse, a diamond panelling was implemented with thin silicone sheets. The silicone, thanks to its elasticity, tolerates the unavoidable imprecisions occurred during relocation and mounting of components. The diamond sheets can be stretched in order to reach the existent points of connection with the wooden structure—the white dots fastened on the tip of the skin holder sticks (Fig. 2.20)—which have a certain dimensional variance considering the virtual model. Having foreseen these possible differences of real production with respect to the virtual model, some stratagems of production have been studied. Despite the use of other tools such as a lathe in addition to the circular saw they would have facilitated the shaping of joint joints, a test was carried out using the saw to dig joints inside the sticks. A series of parallel cuts eliminates a substantial amount of material making it possible to shape a female–male joint (Fig. 2.21a, b, c). For time require-

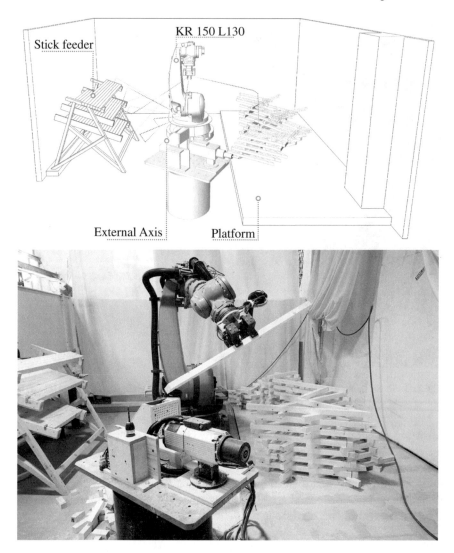

Fig. 2.12 Virtual simulation (up) and real robotic set-up during a production loop (down)

ments, the solution has been discarded, but it is a field of possible development of manufacturing automation.

In order to collect complete documentation on the project, a static analysis of the prototype and 3D scan (Fig. 2.22a) were conducted. The Windmill structural consultant team conducted an analysis of collecting data on the effect of own weight. In particular, there is a vertical lowering and a significant load of the wind effect is assumed once the silicone skin is finished—both positive and negative pressure—with consequent further deformations of the prototype (Fig. 2.22b). The wooden structure has

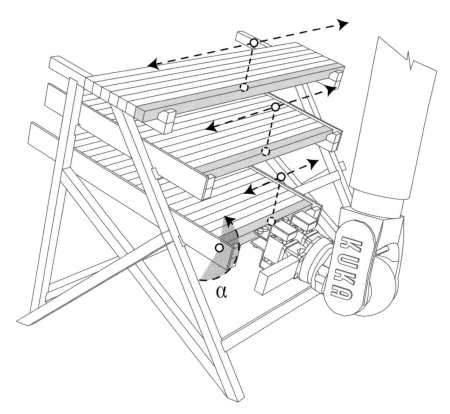

Fig. 2.13 Picking positions vary according to safety tolerances. The angle α shows a rotation while extracting the single stick from the provider

been completed in late February 2016; during the following phases, a silicone skin, transparent to UV radiation, will complement the pavilion as well as the integration of the cultivation system.

2.2.2 Design Workflow for Natural Forks

2.2.2.1 Context

In the context of robotic woodworking, the researches carried out on the topic of natural wood gain relevance after the unprecedented realization of the Wood Chip Barn truss by the AA team based in the Hooke Park Campus (Fig. 2.23). The pioneering project using round wood combined with the precise machination through robotic milling reveals the efficiency of this couple also in an unusual arrangement for timber. The Vierendeel truss is empowered by the reliability of the naturally formed joints

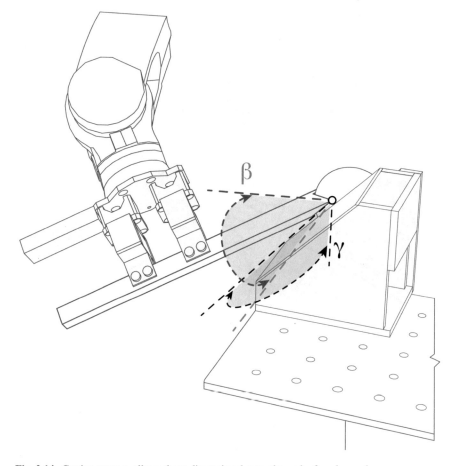

Fig. 2.14 Cutting process allows three-dimensional cuts, shown by β and γ angles

between the stem and the two branches. The set of beech forks, directly harvested from the Hooke forest, has been arranged in its spatial articulation according to meta-heuristic organization logic. This emblematic design experience is in continuity with the ethic of material self-sufficiency developed since the birth of the campus as a site where experimental timber constructions developed. Although with a lower level of scientific rigour, the next case study successfully manages several dimensional tolerances arising from natural wood machination through the conventional tool (i.e. chainsaw) operated with a six-axis robot arm.

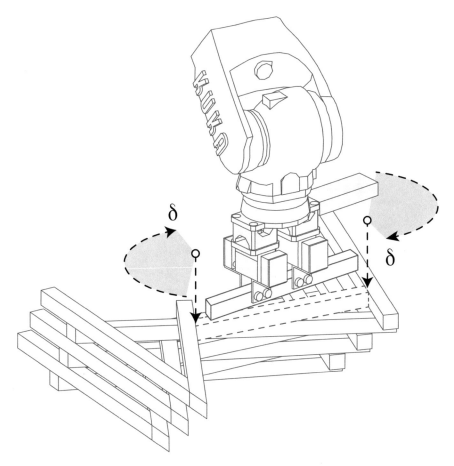

Fig. 2.15 Interlocking positioning

2.2.2.2 Manufacturing Manoeuvres

The Forks Foundry project (Fig. 2.24) is developed during the Chainsaw Chore-ographies[17] AAVisiting School (August 2016) as part of the Robotic Fabrications series of architectural experiments based at the AA's Hooke Park campus. This 10-day long workshop focused its attention on designing possibilities arising from the combination of the versatile world of robotics with traditional techniques. The fruit-ful correlation between craft experience and digital tools enables traditional making techniques to be integrated as design parameters.

---

[17]The 2016 edition had as tutors E. Vercruysse (director), G. Edwards, P. Devadass and Z. Mol-lica (tutors) and as students: E. Azadi, M. Bannwart, J. Blathwayt, A. Carpenter, J. Curry, Ka. Kaewprasert, Ko. Kaewprasert, A. Quartara, E. Rodionov, M. Sharp and C. Thompson.

(a)

(b)

Fig. 2.16 Extreme limits of cutting possibilities

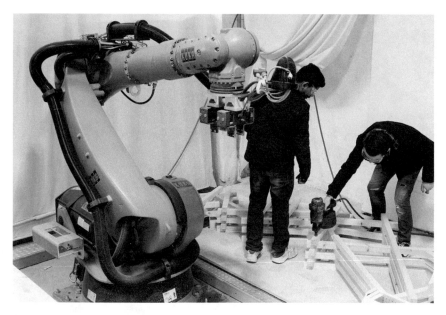

Fig. 2.17 Manual nailing process

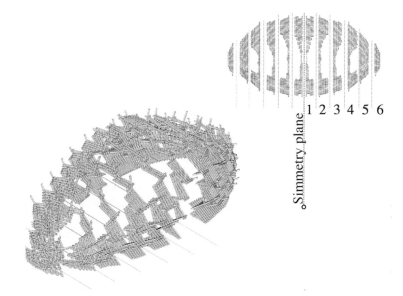

Fig. 2.18 Numbering of different components

The project approach drives non-standard production by means of traditional tools by combining their limitation with precise manoeuvring. The topic of the project was a structure to be erected in the campus that potentially can host the foundry of the park,

Fig. 2.19 Assembly on-site

where students and scholars will have a proper space to carry on physical testing on multiple material connections in particular aluminium-wood joints. Physical models and prototypes have been crucial also for the actual design of the foundry structure.

The structural elements at the disposal of designers have been tree natural forks and rectangular section timber beams (Fig. 2.25). In order to start the designing process, the forks have been digitized and their virtual models have been shaped to become the primary vertical structure. The 3D scan was performed with the help of a photogrammetric processing of about a hundred digital images for each fork. This digitizing process generates dense clouds of points that constitute the three-dimensional virtual representations of the real trunks. The triangulated meshes that are produced through the software incorporate photorealistic textures that allow to check the dimensional tolerances between the virtual model and the real trunk carefully. In order to compare the measurements, on each trunk, a set of three yellow points was painted; knowing the real measurements, they were then compared to the digital ones, finding a very high level of precision. This was also possible thanks to the quality of the photos, taken in an environment as free as possible from obstacles and framing the forks resting on sawhorses (Fig. 2.26). As for the horizontal structure, the wooden beams were considered as solid parallelepipeds with a rectangular section of 150×200 mm and 4900 mm in length.

The joint design was based upon the ancient traditions of Japanese joinery and their transfer into suitable movement patterns for the robot wielding an electric

Fig. 2.20 Assembly test of the silicone skin

chainsaw as end effector. The tool was equipped with a 400-mm-long bar from its body to tip, with a chain thickness of about 5 mm. The different cutting tests carried out revealed that the cut width was 8 mm approximately; therefore, all such measurements have been taken into account during the digital modelling of the joints as dimensional tolerances. Once the general disposition of the vertical and horizontal elements has been determined, the teamwork designed different joints: the connection

(a)   **(b)**

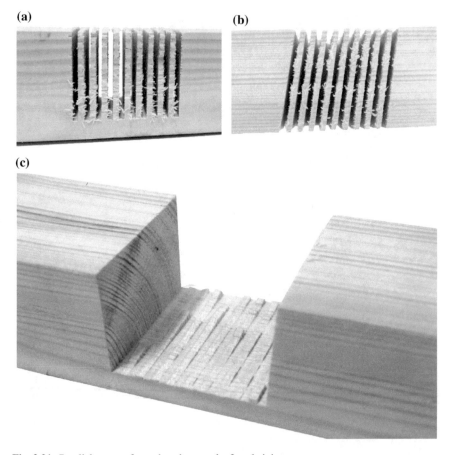

(c)

Fig. 2.21 Parallel cuts performed to shape male–female joints

between the fork's roots and metal L-shaped profiles, drowned in concrete foundation plinths (Fig. 2.27), the forks extremities where the beams rest, and the intersections between beams. Furthermore, the design has considered some parametric joints on their terminations allowing beams extensions at a later stage.

The robot room is equipped with a trolley (on a linear rail) where the wooden elements are placed and they can be easily moved in front of the robotic arm (Fig. 2.28). This versatile set-up helps to increase the robot reach—or work envelope—and reduces the chances of running into singularities[18] or collisions during the cutting

---

[18] Singularities are caused by the inverse kinematics of the robot. When placed at a singularity point, there may be an infinite number of ways for the kinematics to achieve the same tip position of the robot. Six-axis industrial robots suffer from three types of singularities: wrist singularities, shoulder singularities and elbow singularities that can be generally defined as collinear alignments of two or more robot joints that cannot be solved—the robot is locked in position—or can be solved only by instantaneous 180° spin.

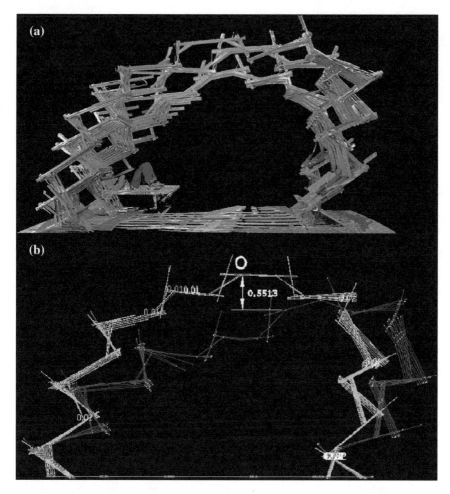

Fig. 2.22 Point cloud model obtained from the 3D scan (**a**), and structural simulation of Sect. 5 (**b**)

choreographies—hence the name of the workshop. The generation of appropriate geometries and the robot tool paths also considered the constraints that the chainsaw forces: all kind of shaping, pocketing or tenon-mortise like connections had to be modelled with straight cuts. Although both the tip and the long side of the chainsaw bar were used to make the cuts, some positions were out of reach and therefore some manual finishing was needed. In particular, some corners have been finished with hammer and chisel (Fig. 2.29). This exercise of translating traditional craft with precise tools was then an experience of how advanced tool tolerances can only be solved by combining digital and analog strategies.

Fig. 2.23 The Wood Chip Barn truss covered with the final roof

2.2.3 Conclusions

Previous renowned design experiences involving computation and digital means of productions have certainly developed higher level of scientific results. In spite of this, the case studies gathered in this paragraph take the credit of experimenting different kinds of tolerances while combining different implications such as the environmental optimization of shape, the mass customization advantages and the integration between technological systems, low-engineered set-up and natural material. The exploration of computational software and robot machining strategies allow and assist architects with the generation of innovative spatial materializations.

Customization has been crucial in both the design experiences and it also concerned the hacking of standard woodworking tools placed on the rotary table, such as the mitre saw which allows three-dimensional angled cuts, or the pneumatic gripper fixed on the robot flange used as end effector, and also the sticks supplier which provides wooden profiles in three different lengths in order to reduce waste material or simply a chainsaw carefully manoeuvred in space. It is really about understanding how to design and build your own tools.

Future improvements, apart from tool upgrades, should move forward the first robotic construction experiments that adopted logic of automation directly from the

Fig. 2.24   The Forks Foundry project combining natural wood and timber beams

Fig. 2.25   One natural fork and timber beams on the background

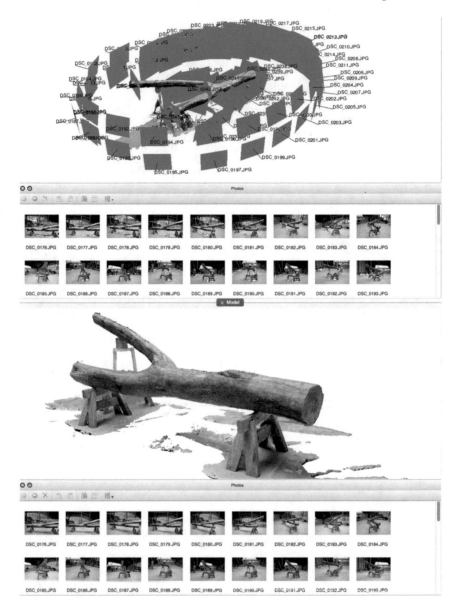

Fig. 2.26 Photogrammetric scan process generates a 3D model by repositioning images

industry. The real potential that digital technologies are carrying is to overcome the traditional way of fabrication; instead of trying to replicate industrial procedures, architects should focus their design on complementary processes that were impossible without current technologies.

Fig. 2.27 Metal profiles, drowned in concrete foundation plinths

2.3 Material Enhancement

The following chapter investigates novel performative building systems based on strip topologies applied in timber construction and the integration of enhanced material properties in the computational design process. It provides an introduction to strip patterns, elastic bending and wood hygroscopicity with the acknowledgement of such concepts for strip planking in boat building, and postulates how the technique's principles can be translated to architecture. The multi-scalar strip patterns, together with the lightweight qualities and mould-less assembly of double-curved geometries, reveal a great potential for buildings and structures. These principles are introduced through prototypical architecture built over the last decade, giving an overview of different approaches.

Next, a series of study cases showcase these novel applications in more detail, explaining computational design methods and material integration. The strip active bending method proposes assembly processes without falsework, faster and more affordable than common single-face panelling systems. The scalability of the strip pattern allows the design and construction of artefacts ranging from small paper models to large-scale structures. This approach addresses to complex curved structural surfaces constructed from discrete elements, exploring structural optimization techniques through dynamic relaxation and particle-spring systems.

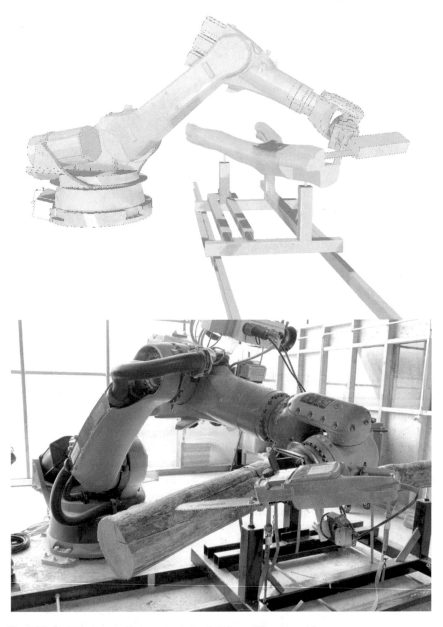

Fig. 2.28 Robotic set-up allows to reach the fork from different positions

Structurally, double-curved, ultrathin timber geometries with self-supporting capacity showcase increased payload qualities achieved through the bending active principle.

Fig. 2.29 Particular corners to be finished by hand

Research explorations are tightly connected to the material properties of extremely thin plywood, hygroscopicity and bending technique; they are combined together to evaluate how the elastic properties and flexibility can be exploited in favour of the global mechanical and structural characteristics. The design strategy is not necessarily connected to the discretization of a given surface. On the contrary, the end result is primarily influenced by the materiality of the strip morphology and aims to explore and optimize the structural performance and stability through the active bending principle. Furthermore, the research is mainly oriented to the augmentation of material properties, specifically on the hygroscopicity of timber, which is applied through the temporary incremental manipulation of moisture content in order to enhance the bending capacity of the material. Moreover, the lamination of timber composites is presented through multilayered and differentiated material orientation that results in improved strength and stability.

The final topics discuss the fabrication-parallel-construction process integrated in the seamless design-to-construction chain. The on-site fabrication approach bridges all team members, design, fabrication and construction for adaptable collaborative work. Thanks to numerous advantages praising structural performance, design freedom, fabrication and assembly processes, the conclusions show great potential for the application of strips in timber construction.

2.3.1 Strips, Elastic Bending and Hygroscopicity

2.3.1.1 Strip Planking in Boat Construction

Strip planking systems have been developed and improved for more than a century; while it has a long history in wooden boat construction, today's advancements in naval architecture have been made possible by the great progress of timber composites. Wood is the traditional boat building material, continually preferred to modern material for its accessibility and cost; it is very strong, lightweight and is not time-intensive or fatiguing to work with. When comparing its strength against weight, wood is stronger than steel, most fibreglass, and aluminium. Strip-building allows the reproduction of any double-curved geometry at any scale of boats. This method is accessible, the equipment is minimal and the technique is very tolerant of mistakes. The simplest and oldest planking technique called clinker-built or lapstrake, a common boat building method system where the edges of hull[19] planks overlap. On the other hand, carvel construction is a method where plank edges are butted smoothly seam to seam. This method began with the introduction of waterproof glues and the development of bead-and-cove profiles, routed on the strips' edges. The common assembly of strips around double-curved structures requires connecting the elements without gaps, and bead-and-cove profiles are often chosen for this insignificant interstice. A less common boat building technique, stitch and glue, stitches[20] together plywood panels and glues the gaps with resin, eliminating the need for frames and ribs. This construction system uses a relatively small number of wood pieces and a boat can be built in strikingly short time. The development of speed strip planking system, currently used by Maritime Wood Products, was a notable advancement in strip assembly. "Speed Strip represents a significant step beyond the bead-and-cove concept. Each strip is now tongue-and-groove in cross-section. The advantages offered by this system include: a 'snap-fit', which reduces the number of fastenings; a reduced number of mould stations and frames; reduced glue squeeze-out; and improved fairness, both inside and out".[21] The time saved in the process is greatly improved over earlier methods (Fig. 2.30). For these systems, several edge connecting variations combining edge profiles with gluing, stitching and nailing were developed (Fig. 2.31).

Nowadays boats aim to be light, strong, rigid and waterproof and achieve properties akin to a monocoque fibreglass shell more consistent in their integrity than to a carvel-planked boat with many ribs. The development of new adhesives with higher performance such as epoxy allows connection and gluing of edge-to-edge wooden

---

[19]The hull is the body of a boat.

[20]The stitches are usually done with copper wire.

[21]Paul Lazarus, "Improving The Efficiency of Strip—planked Construction: new 'speed strip' system cuts time and labor—by a lot", issue of Professional BoatBuilder, reprinted from the April/May 1997.

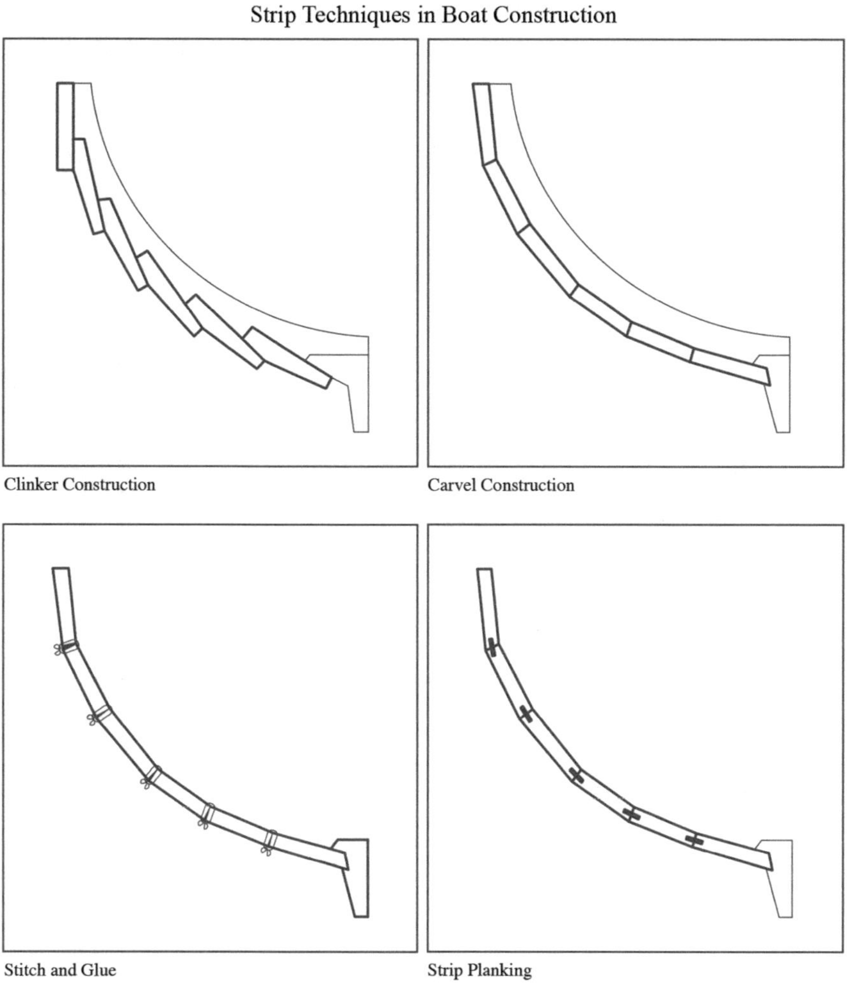

Strip Techniques in Boat Construction

Clinker Construction

Carvel Construction

Stitch and Glue

Strip Planking

Fig. 2.30 Strip techniques in boat construction, diagram by D. Stanojevic, 2018

strips over a temporary frame[22] with no need for fasteners such as screws and dowels. Once faired[23] and clothed with fibreglass, these thin, composite, wooden shells became rigid, strong and waterproof giving a new modern vision for strip construction.

---

[22]Forms are temporary moulds, acting like a skeleton ensuring a supporting surface (usually a plywood sheet). Once the outside of the hull has been glued, glassed, and is rigid enough, the mould can be removed.

[23]Fairness is used to describe if the surface of the hull is fair or smooth; when an unfired surface has bumps or hollows, needs to be fired or smoothed out.

Strip Techniques in Boat Construction

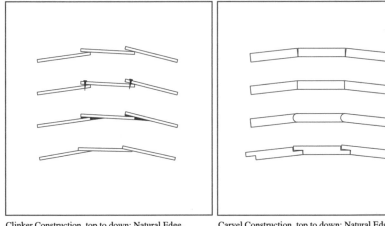

Clinker Construction, top to down: Natural Edge,   Carvel Construction, top to down: Natural Edge, Planed
Nailing, Glueing, Scarf Joint and Glue.   Edge, Bead and Cove, Lapship Edge.

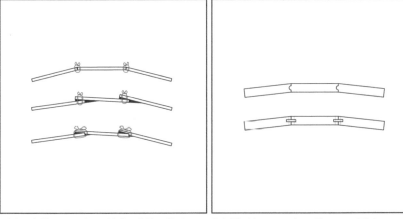

Stitch and Glue, top to down:Stitch and Glue, Common   Strip Planking, top to down: Tongue and Groove,
Clinker Stitch and Glue, Lap and Stitch.   Mortise and Tenon with Dowels.

Fig. 2.31 Joint variations for strip techniques in boat construction, diagram by D. Stanojevic, 2018

In boat construction, cedar is the most common wood utilized among the many types available on market. The wood is chosen by its bending capacity and strength, and in case of necessity, the epoxy can supply supplementary strength. When the strips are not long enough to cover the hull from an edge to another, a scarf joint is introduced to extend the strip with several members. To assure a homogenous bending, the join length should be at least around eight times the strip thickness. Wood bending has limits; it is often necessary to pre-dampen for improved flexibility, and an excellent method to treat planks is with a streaming box. The wood strips have to be thin enough to bend around the shape of the hull so that the produced boats are ultralight and stiff. To attach the strip to the frame, the easiest and fastest way is to use staples or nails, which are removed when it is time to fair the boat. Whichever method is chosen to plank, there will always be an area that will require trimming and fitting of strips. After fairing and cleaning, fibreglass cloth is applied. There are very thin fibreglass types, which add almost no weight to the boat. During the layup of the cloth, it is important to take care to not create bubbles; it takes two or three coats of epoxy to get the coating smooth, and finally sanding is needed to achieve a perfectly polished surface before varnishing. The varnish is applied with ultraviolet protection to safeguard the epoxy from sunlight.

The hull can be built with a differentiated strip pattern, for example, with different profiles on port and starboard.[24] Levels of strip hierarchization can be applied through techniques such as cold-moulding. This method is implemented through lamination; it uses two or more layers of thin wood oriented in different directions. The matrix can have a base layer of strip planking followed by multiple veneer sheets.

Strip construction is a relatively simple way of developing new shapes, it allows designers to experiment on more complex geometries, work on different scales and yet to keep the structure both lightweight and stiff. Based on these principles and the analysed strip methods, it is noticeable how both archaic and new advancements of planking techniques for wooden boats showcase great potential that can be applied to the field of contemporary strip construction systems in architecture.

2.3.1.2 Architectural State of the Art

From a spatial point of view, strip systems can be used to represent three-dimensional objects defining space, form and structure simultaneously, while "in computational design, material elements can be defined by behaviour rather than shape", arising new performative potential through the material driven design approach (Menges 2015).

While strip discretization has existed since a long time in industries such as aeronautics, marine, inflatables and tenso-structures, strip morphology has been introduced in architecture by M. Fornes, who has built a great number of large-scale installations and prototypical architectures based on the development of strips at Theverymany studio. The principle of strip morphology was initiated in order to simplify and reduce assembly time from the previous paradigm of panelling based on singular planar face.

Applications of strip morphology on material driven design approach have been widely explored at the Institute for Computational Design and Construction (ICD) and at the Institute of Building Structures and Structural Design (ITKE). In their Research Pavilion 2010, the active bending principle demonstrated an alternative approach to computational design where the generation of form is directly influenced by the characteristics of the material.

J. Lienhard, one of the pioneers of this method, developed finite element simulation strategies to form-find such structures. Bending active structures are

---

[24]Port and starboard are, respectively, the left and the right side of a boat.

Fig. 2.32 Bend9, R. La Magna

currently a focus of R. La Magna and S. Schleicher's research, who are challenging material properties through finite element methods investigating both bottom-up and top-down design approaches. Their investigation on bending active plate structures demonstrates how to translate computationally elastic bending through nonlinear analysis and to form-find the global geometry as a result of equilibrium of embedded forces. "Moreover, it is particularly interesting that by discretizing complex, double-curved geometries with flat-produced, single-curved panels, one can avoid the high costs that usually relate to the manufacturing of complicated moulds. Compared to this, bending active structures can be fabricated quickly and cost-efficiently since they require little or no formwork at all".[25] The project Bend9 by R. La Magna (Fig. 2.32) is particularly relevant to this research as it explores the top-down design approach for active bending plates and further integrated a double-layered building system. Along with this project, the Berkeley Weave project by S. Schleicher and R. La Magna (Fig. 2.33) is another demonstrator of the active bending system applied to weaving techniques where the computation of the twisted elements is evaluated structurally on both global and local level.

At the EmTech master programme at the Architectural Association School of Architecture in London, under the lead of M. Weinstock, there are two projects

[25]Schleicher S, La Magna R, Zabel J (2017) Bending-active Sandwich Shells Studio One Research Pavilion 2017. In: ACADIA 2017 Disciplines & Disruption, Proceedings of the 37th Annual Conference of the Association for Computer Aided Design in Architecture (ACADIA), Cambridge, pp 544–551.

Fig. 2.33 The Berkeley Weave, S. Schleicher and R. La Magna

designed strategically through active bending which deserve to be mentioned. The first, the AA/ETH Pavilion, was part of a collaboration with the DARCH Chair of Structural Design from ETH Zurich during 2011–2012. The pavilion was a temporary timber lightweight construction designed and built based on the material's bending behaviour and the fibre direction, which explored how strategic cuts within the sheets influence the bending capacity. "The structure of the pavilion is based on three bent panels of plywood with non-standard dimensions, up to 2.3 m in width and 10.3 m in length" (D'Acunto and Kotnik 2013). The second project, The Twist, realized in 2015, developed a building system where sinusoidal strips combined with straight members achieved self-supporting properties. A lightweight surface was realized through the study of the assembly and locking solutions integrating the connection of twisted elements.

At the IBOIS in Lausanne, the laboratory for timber construction of the EPFL is working on the Timber Fabric project, which is investigating the principle of fabric and basket weaving methods, exploring characteristics such as the integrity of the whole governed by the friction between yarns. Keeping in mind the yarn's length in textile can be extended with no limits, this system can be translated to contemporary architecture and provide the possibility to build timber structures of infinite length.

At the ICD, the performative wood research line collects a series of wood projects with strip morphologies and elastic bending focused on exploiting the formal potential of structural systems defined by the join itself. An approach developed on this notion is the ITECH M.Sc. 2016 Thesis Project Tailored Structures by M. Alvarez and E. Martínez which explores new ways of fabricating wood shells through the industrial sewing techniques, where sewing is proposed as a new construction joint with comparable strength to gluing methods and standard fasteners. This research has been developed further with the ICD Sewn Timber Shell 2017 integrating a triple-layered system and showcasing how the adaptive robotic fabrication allows the scaling up of the structure and new design opportunities for spatial articulation.

Another interesting proposal by the ICD was the integration of permanent form-works in order to accommodate strip elements on double-curved surfaces. The first example is the Robot-Assisted Assembly in Wood Construction 2015, which relies on a robot assistant to fabricate the substructure, where the strips with pre-drilled screw holes allow an intuitive assembly on-site. Later in 2017 came the Robot Made: Large-Scale Robotic Timber Fabrication in Architecture, where the ICD fabricated a double-curved wall with a multi-axis milling set-up, fabricating a strip facade on a diagrid substructure.

One of the earliest ICD projects referring to a pure bending through assembly technique with joint locking system is the Diploma 2012: Connecting Intelligence by O. D. Krieg, which develops novel assembly logics through elastic bending and finger joints to lock strip members and consolidate them elegantly and cleanly with wood glue. More recently, the ITECH M.Sc. Thesis 2015 "Architectural Potentials of Elastically Bent Segmented Shells" by G. Kazlachev introduces to strip elastic bending a double-layered system and a quick assembly method through a combination of finger joint with zip ties.

Additionally, the ICD's research explores the hygroscopic properties of wood and the design of material's active and reactive properties, as exhibited in the Hygro-scope project: a meteorosensitive installation that responds with movement to climate changes. Based on this background, the ITECH M.Sc. 2015 Thesis Informed Connectivity by D. Stanojevic explores the hygroscopic potential of wood as an alternative to common fasteners focusing on applying mono-material and assembly logics to interlock cellular three-dimensional plates with the natural wood expansion. Continuing with this topic in the Robot Made: Double-Layered Elastic Bending for Large-Scale Folded Plate Structures 2016, the project showcases the full potential of the combination between elastic bending of double-layered systems, finger joints and hygroscopically actuated fasteners.

2.3.1.3 Study Cases—Design Approach

In the last 2 years, for the research on active bending lightweight structures carried on by D. Stanojevic with the collaboration of G. Kazlachev, three study cases have been developed and tested with full-scale prototypes, The Woven Wood, The Synthesis of Strip Pattern and the Laminate Pavilion. The scientific development and digital fabrication methods were investigated through computational design approaches and material tests, defining buildings systems and design space possibilities. The results have been brought to a series of workshops and open to participants who worked on design, fabrication and construction of such systems. The three projects "explore how the bending behaviour of wood can be integrated as a generative driver

Fig. 2.34 Woven Wood, D. Stanojevic and G. Kazlachev with Noumena, at Sbodio32, designed for Milan Design Week, Milano, 2016

in computational design, and how today wood can be physically programmed to perform more variable and differentiated bending figures through additional, digitally controlled fabrication techniques" (Menges 2012). Moreover, the versatility of the systems is well suited for construction techniques "to form complex, lightweight structures from initially simple, planar building elements" (Menges 2012).

The Woven Wood develops a novel material treatment and fabrication process by investigating the relationship between the performance of flexible timber structures and basket weaving methods (Fig. 2.34). This project bases its system on the ICD/ITKE Research Pavilion 2010, using interlaced wood strips to generate a self-supporting lightweight structure. Another important reference is the project Berkeley Weave by S. Schleicher and R. La Magna: the structure introduces new design potential by integrating both bending and torsion of slender strips into the design process. The additional target in the Woven Wood was to achieve a double-curved self-supported wall with differentiated thickness, where the strips gain strength through their single curvature in opposite directions combined with local torsion forces. The design development implemented a top-down form-finding approach, which took advantage of algorithms to relax and define the final buildable geometry through the limits of the unrollable curvature. The fabrication process went from CNC-machining planar sheets, wetting wood strips to increase flexibility and assembling by interlocking the pre-programmed slits.

The Synthesis of Strip Pattern explored off-grid column structures based on a ring-like topology, forming a continuous, smooth surface, minimizing the quantity of

Fig. 2.35 Synthesis of Strip Pattern, D. Stanojevic and G. Kazlachev with Noumena, at Nodo and IAAC, Barcelona, 2017

material by adopting slim face-to-face edgewise connections, with a quick and simple assembly method assembled by a small team (Fig. 2.35). A bottom-up approach has been chosen, where the overall design has been entirely form-found, controlling and grouping topological, internal stress forces and constrained only by the predefined boundary conditions. The laser cutting patterns of thin plywood sheets integrated both linear and lateral boltholes and it took place in parallel with the wetting process, which lead to a more flexible assembly. The built structure was designed to be disassemblable and after a series of exhibitions and relocations, it is currently part of a long-period exhibition at the Institute for Advanced Architecture of Catalonia in Barcelona.

The Laminate Pavilion explores mould-less lamination through bent timber assembly for segmented double-curved shells in order to build without supports and to gain strength through a multidirectional and multilayered material composite (Fig. 2.36). Pure, structural expression through form was limiting the habitable interior space, which led to return to top-down design explorations. The design investigation explored two distinct curvatures, synclastic and anticlastic. Although the anticlastic typology had exhibited higher potential to be load bearing, the synclastic curvature was chosen in order to direct the shell design as to host as many people as possible in its interior, and, at the same time, to create a comfortable spatial experience, separate from the external environment. The project covered an entire year of research, which integrated form-finding design technique, double-sided milling strategies, wood steaming, bending assembly and composite lamination.

Fig. 2.36 The Laminate Pavilion, D. Stanojevic, at Cedim, Monterrey, 2017

"For the future development of bending active plate structures, it may be desirable to prioritise a top-down approach, which gives more weight to the target geometry and thus more freedom to the designer" (La Magna et al. 2016). Moreover, the active bending approach should be tested on thicker wood sections in order to scale up the system to large span structures.

2.3.2 Material Calibration

2.3.2.1 Strip Active Bending

With the contemporary development of computational tools in architectural design, it is becoming much easier to discretize complex geometries into simple elements, as well as to generate complex structures out of units with defined characteristics.

The assembly of strips is faster and a more affordable option when compared to common systems for building double-curved structures. Additionally, the system has a high potential to be applied in construction with no need for additional, temporary formwork. The design strategies can either follow linear or nonlinear approaches; in fact, the strips can be embedded in the global geometry before executing form-finding operations, or they can be extracted through mesh subdivision from geometric analysis, mesh faces properties, or algorithmic behaviours. The size and the amount of strips define the resolution of the curvature and subsequently the number of unique

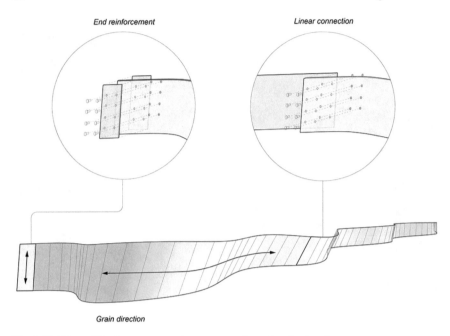

End reinforcement   Linear connection

Grain direction

Fig. 2.37 Woven Wood, strip connection system and wood grain direction

pieces and their relative connections. The strip topology sets the rules process of assembly, which directly informs the design at both local and global levels.

The assembly sequence and strip orientation influence the design opportunities, not only digitally, but also physically because of the bending assembly degrees of freedom. The active bending principle counts on the material properties to describe the geometrical design space, designating the minimum radii achievable by the strip dimensions and the material elastic bending capability. In all the three study cases, a series of bending tests were conducted with different strip dimensions and wood species, where the proportion between these two aspects defined a factor, which was then integrated in the design code, delineating a range of bending limits.

In Woven Wood, in order to guarantee structural continuity, the sub-strips were joined in a larger strip through fastening overlapped edges (Fig. 2.37), which were then locally bent and fitted in slits in order to interlock the strips (Fig. 2.38). Each slit width was designed to fit the neighbour strip with minimal tolerance, informed by the incidence angle of the two members and the material thickness. The restrictions of this locking method applied on long strips required a high amount of people to coordinate and execute the assembly simultaneously. The building system integrated wood cylinders to act as compression bars against the strips, designed within a range of maximum amplitude; these elements pushed away the strips one from one another, increasing local stiffness (Fig. 2.39).

In the Synthesis of Strip Pattern, the sub-strips were assembled into larger strips through additional plates and bent into a ring component to achieve self-stability,

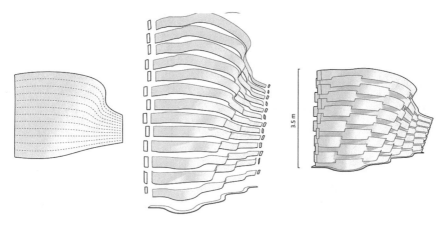

Fig. 2.38 Woven Wood, surface division—Strip generation, joining system integration

which rotates slightly outwards in a cone-like shape in order to accommodate one ring into other through tangential connections.[26] This project aimed to reduce the assembly team to a group of only three people, making it simpler and faster.

In the Laminate Pavilion, the assembly logic required the integration of joint systems to connect strips segments into shell strips in a linear fashion and a solution to attach each strip laterally and edgewise to each other, maintaining surface continuity. Both cases were solved with a lap joint and staple gun assembly[27] (Fig. 2.40). The shell strip is anchored on one end to the shell base, and then gradually connected edgewise to the previously mounted strip, and finally, the other strip end is also locked to the shell base. The Laminate Pavilion was almost entirely assembled by only three people.

Both the Synthesis of Strip Pattern and the Laminate Pavilion required the simultaneous assembly of both strip sides. In the ring-like topology, the two faces were easy to reach at once, while the arches of the last project utilized a temporary bridge platform to access the outer shell to speed up the process.

2.3.2.2 Hygroscopic Enhancement

Wood is a natural material with intelligent, ingrained properties; once bent it tries to return to its original shape unless fixed and when assembled through bending, it is stronger. In this research, the hygroscopic properties are pushed beyond the material characteristics, exploring the enhancement of bending through the material's increasing level of moisture content.

---

[26]The overlapping system recalls the clinker construction technique for boats.

[27]Analogy to the scarf joint adopted in boat construction. In this project, the length of the linear lap joint was set to 25 times the strip thickness.

Fig. 2.39 Woven Wood, compression cylinders

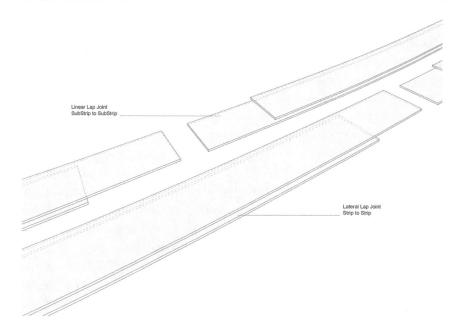

Linear Lap Joint
SubStrip to SubStrip

Lateral Lap Joint
Strip to Strip

Fig. 2.40 The Laminate Pavilion, linear and lateral connections

Under this treatment, the bending radii limit can decrease by 50% of the material's dry state, expanding the design possibilities beyond the conventional state of wood. Other advantages are a flexible assembly with less springback of the strip, an easier manipulation of the material on-site and stiffer material properties due to internal stress forces.

This process has been tested with steaming and soaking different types of plywood in water, with controlled relative humidity, temperature and period of exposure. The bending tests effectuated before and after the treatment showed a great difference in behaviour (Figs. 2.41 and 2.42).

Although water is considered to be an enemy of wood, it is important to clarify that there are species prone to deteriorate and others to gain qualities once in contact, therefore the time of exposure should be tested and adjusted to the species' reactivity to water.

While working with plywood, some characteristics when dealing with humidity must be avoided:

- Water saturation due to excessive exposure, which might create internal air bubbles, affect the material with plastic deformation and introduce growth of mould.
- Poor bounding can cause delamination.
- Uneven distribution of moisture might showcase unpredictable rigidities in confined zones.

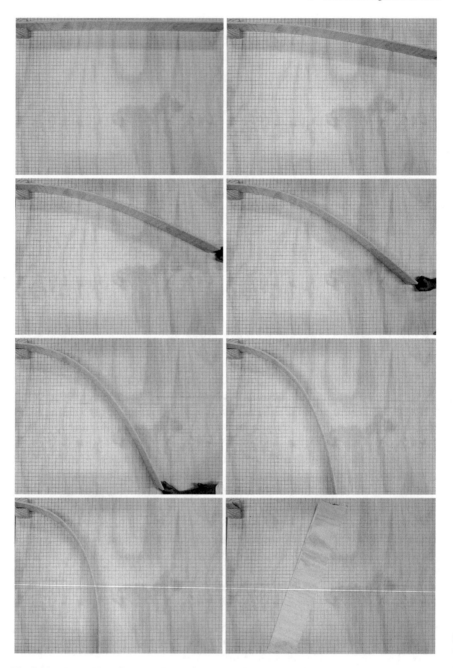

Fig. 2.41 Plywood bending tests, 9 mm dry plywood, Cedim, 2018

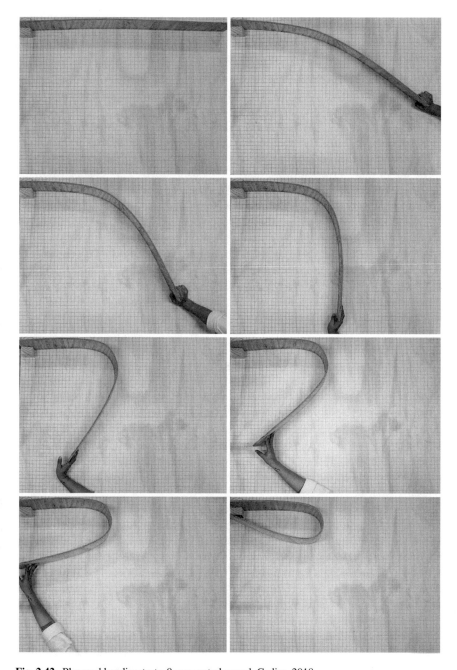

Fig. 2.42 Plywood bending tests, 9 mm wet plywood, Cedim, 2018

- The environmental humidity can be also a critical aspect in construction, sun or heat exposure can speed up the drying, therefore it is necessary to keep the wood moist especially for the outdoor assembly.

Humidity monitoring becomes not only a concern for the material, but for the environment as well, the temporary augmented flexibility demands a precisely scheduled time to wet the material and prepare it for the assembly.

Dimensional change of wood can be calculated in order to prevent damage in the structure, and integrated through relative tolerances in the connection design.

Once the plywood has been dampened to its maximum capability, it is crucial to utilize the element before it dries; once the moisture percentage re-equalize to the environmental dry temperature, the material loses a part of the initial flexibility and a second treatment will most likely cause damage to the wood grain.

The Woven Wood and the Synthesis of Strip Pattern adopted a material thickness of 3 mm and the soaking process, where birch plywood was chosen because of its better bending and resistance properties compared to other plywood types available on European market. The Laminate Pavilion base its system on steaming of 9 mm thickness plywood, where pine plywood was selected as the most available material in the region of Nueva Leon, Mexico, as it had good load-bearing resistance for the design intentions.

2.3.2.3   Timber Composite Construction

Wood is, by nature, a composite material with anisotropic properties made of cellulose fibres and lignin, considerably stable along its longitudinal direction, while quite flexible parallel with the grain.

To achieve the maximum strength from a flexible material, the Woven Wood project applied grain direction along the strips. On the contrary, The Synthesis of Strip Pattern utilized strips with grain perpendicular to the strip direction, in order to reach higher flexibility.

Plywood is an engineered material with alternating layers of grain direction, which controls its strength but also retains a degree of flexibility. As seen through hygroscopic programming, this material can increase its flexural limits, but the strength remains related to the wood thickness, or in other words, to the quantity and density of fibres. This aspect can be addressed in the lamination processes: a laminate is a permanently assembled object by overlaying substrates, creating an extra durable body, welded together in its structural integrity.

The Laminate Pavilion construction system implements a composite lamination on-site (Figs. 2.43 and 2.44), where the individual layers consist of wood strips, aligned with the directions of the topological matrix, and a fibreglass reinforcement is introduced to improve the tensile strength and bending stiffness of the laminate as a whole.[28] Fibreglass layers have been sandwiched between plywood sheets (Figs. 2.45

---

[28]Race boats or high-end performance cruisers are built through fibreglass lamination to reduce weight and maximize strength. Alternating layers direction increases the laminates resistance. Each

Fig. 2.43 The Laminate Pavilion, on-site lamination process, plywood–fibreglass composite

and 2.46) and tested on multiple scales, where strength and stiffness of the specimens were measured by comparing the reinforced samples with those that were not modified. The implemented performances were validated through composite arch samples with a sequential addition of weight until breakage occurred. The experimental results highlighted that the integration of the fibreglass layer increased the material resistance by 50–60%.

The simultaneous hand-layup[29] of the resin-impregnated fibreglass and wood layering through an open moulding method facilitated the fibre bonding between the two materials and avoided the formation of voids in the composite.[30] The pressing process was facilitated by the segmentation of the strips in the design stage, which allowed the elements to anchor easily, without a significant springback.

Each layer direction was designed to accomplish its anchoring function, consolidating structurally, keeping elasticity and preventing wood from splitting. It is relevant to state that in case of static overloading, because of the given assembly

---

ply can align its coordinates axes with the material matrix, modulating material properties as such as orienting the principal directions of elasticity.

[29]The simplest method to layup fibreglass is to apply the fibres with a brush, impregnated from a bucket of catalysed resin. Another method is through special rollers, which contain catalysed resin directly in the roller's head, but the most advanced and practical technique is with spray guns, where the resin and catalyst are mixed at the moment of being sprayed.

[30]During the layup, there is the risk that small pockets of air, called voids, get trapped between the layers. They promote blistering by providing locations for moisture and therefore reduce the laminate's integrity.

Fig. 2.44 The Laminate Pavilion, on-site lamination process, stapling plywood sheets

Fig. 2.45 The Laminate Pavilion, plywood–fibreglass sandwich system, open moulding method

system, the laminated wood can potentially fail with a ductile compression failure. "Ductility is the extent to which material can plastically deform without losing its

Fig. 2.46   The Laminate Pavilion, plywood–fibreglass sandwich system, lightweight application

load-bearing capacity. Ductility is essential in accidental situations, areas of high seismicity, and in cases of static overloading. In general, ductility can be understood as a safeguard against the unknown".[31] This characteristic is particularly relevant for lightweight structures as they have higher potential to common massive buildings methods to adapt to foundation systems, which responds to telluric vacillations.

2.3.3 Computational Methodology

2.3.3.1 From Strip to Surface Directionality

The design proposal generated by the optimization algorithm does not yet produce a structure, but leaves us with a geometry that needs to be interpreted and translated into a constructible shell.

(I. Lochner-Aldinger and A. Schumacher 2014)

Computationally, the topology of a surface is defined by two directions, marked by UV coordinates system. While working with meshes, it is possible to reconfigure the directionality through mesh subdivision. For instance, a quadrilateral mesh has two directions, while a mesh with a hexagonal structure has three. With re-meshing algorithms, it is also possible to reconfigure a mesh grid by behaviour, for instance, the mesh machine component for Grasshopper developed by D. Piker can reorganize the triangular mesh faces with dynamic bottom-up mesh topology optimization. This approach can be powerful for complex geometry discretization into irregular strips, but the individual elements might not be adequate for wood strip application. The strip elements extracted from such re-meshing algorithms are too segmented for wood application; because of the grain orientation, the strip profiles should be as linear as possible. Therefore, the design of strips patterns for timber structures should maintain a simple quad-based topology with constant node valence.

In order to include geometric features as strip developability and customizable mesh resolution in the process, one must begin by topologically subdividing quad meshes in one direction, according to the amount of strips, and to the other direction, according to the strip resolution. The amount of strips is directly related to the strip width, which is chosen by the maximum material bending capacity. The thinner the strip, the easier the strip bends. The drawback of having too many strips can be increased cutting and assembly time, and it is therefore necessary to find a balance between design and production. The generation of the topology does not integrate forces at this time, but considers the forces' behavioural consequences. In some cases, with a top-down approach, it might be preferable to begin by lofting surfaces from imposed curves by the user. In geometrical form-finding, it is necessary to convert the initial surface to a mesh, described by links and vertices, and utilize these objects as inputs for particle-spring systems and mesh relaxation. "Quadrilaterals in a quad

---

[31] Pirinen M, Ductility of Wood and Wood Members Connected with Mechanical Fasteners, Master's thesis AALTO UNIVERSITY School of Engineering, Degree Program in Structural Engineering, Espoo. 2014.

Fig. 2.47 Synthesis of Strip Pattern, surface directionality and its curvature transition

mesh in general are not planar"(Pottmann et al. 2007), and as three-dimensional mesh relaxation deviates the coincidence of the face vertices from the same plane, in order to target developability, mesh relaxation integrates mesh face planarization among the design goals. When mesh relaxation takes action, the process consists on an iterative principle, which keep adapting the geometry according to the internal forces, planarizing individually the mesh faces, and external ones, adapting to boundary conditions and gravity. The geometry might not definitely get to a stable shape but is usually lowering the oscillations getting as close as possible to the most balanced solution. Strip direction delineates the surface direction, which describes the topological aspects. More complex situations with mesh intersections require additional topological rules such as different vertex valence, constraining vertex motion and fixing a range of angles between links to allow transition from one mesh to another.

A strip-originated surface directionality manifests its appearance on the global level as texture, embedding new aesthetic qualities (Fig. 2.47). The flow of strips creates a visual perception of continuity and smooth transitions of unique elements with uniform design rules.

In Woven Wood, the top-down approach began with an input surface based on space and exhibition requirements, while mesh subdivision was informed by mesh relaxation of the woven strips.

For the Synthesis of Strip Pattern, the bottom-up strategy defined the amount of strips directly from the initial mesh. In addition, the strip width was designed as a gradient from wider at the base to thinner at the top, allowing a more optimized curvature resolution on the transition through the column intersections. The maximum

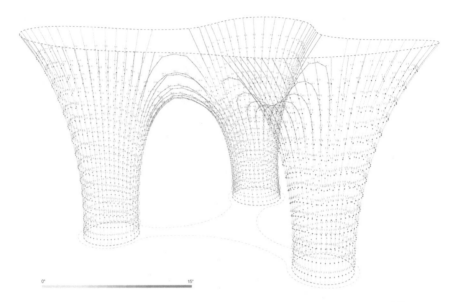

Fig. 2.48 Synthesis of Strip Pattern, strip inclination angle, the maximum was set to 15°

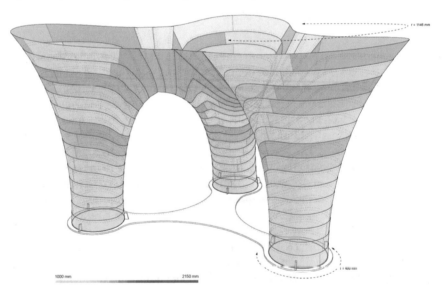

Fig. 2.49 Synthesis of Strip Pattern, strip length, from a minimum of 1000 mm to a maximum of 2150 mm

angle of the strips' tangential connection was of 15° (Fig. 2.48). The strips were designed out of 2–4 sub-strips joined by a plate, where the sub-strip length varied from 1 to 2.15 m (Fig. 2.49).

The synclastic double curvature of the Laminate Pavilion was defined with a top-down approach. The discretization of the shell in developable strips began with quite few complications; a double-curved geometry visualized through a generic quad mesh with naked edges fixed to a predefined boundary does not allow much flexibility for planarization. The strategy adopted was a differential relaxation, which allowed one direction to have more freedom against the other. Conducted through goal optimization to planarize singular quads units of the strips, this computational method extended the limited degree of freedom for planarization. The Laminate Pavilion bases its topology and resolution simultaneously on the two mesh directions. The first direction is used to define the amount of strips of the first layer, and the second to define the strip number in the opposite direction of the additional layer.

The topology has the quality to be the DNA of an object and therefore to define the force behaviour on relaxation and optimization processes. Design with topological assignation can be applied to form-find geometries with aspects of developability as well as to approximate the input geometry into newly formed developable elements. The vertex valence in the strip projects applied for timber structures is kept to a degree of four, and only in the Synthesis of Strip Pattern, the meeting point between the three columns has a degree of five. A high valence means slim elements with narrow angles, where the connections become much harder to fit.

2.3.3.2 Code Bridges

The contemporary computational design process bases its development on the integration of a multitude of parameters which, in order to produce a file-to-factory definition, are structured upon different steps from design to the final construction. The structure of the code can be summarized in three wide stages: global design, construction system and production output. In the context of the presented projects, the global design occupied less than a quarter the full code definition. The entire workflow of the three phases is subdivided in: 1.0 strip topology generation, 2.0 mesh relaxation, 3.0 building system, 4.0 strip subdivision, 5.0 connections, 6.0 unrolling and 7.0 nesting. Every step has subgroups and different levels of complexity, where the importance lies in the input/output fixed structure with proper code labelling.

During the code development, until the last stage, the level of parameter resolution such as mesh subdivision, quantity of strips and preview settings was kept low, as it allows a faster processing ratio of data and a quicker opportunity in generating geometry iterations. However, for prototype testing and final geometry definition, it is necessary to increase the resolution, in order to precisely evaluate the geometrical quality, the joint details and assembly tolerance. This leads to an exponential growth of data to process. Long codes with extremely high computational data to analyse were divided into three separate codes, corresponding to the three stages mentioned above. This method of code distribution was assigned to different teams, which were developing the stage codes in parallel, while maintaining identical input/output data structure.

Codebreaking or code-splitting communication relies on establishing code bridges by topological means. It is a considerable advantage to retain the tree structure of the code as hierarchical as possible but also to filter specific elements of the structure into topological categories such as base connections, boundary elements, intersecting areas, etc. Grouping indices instead of elements is a powerful approach, for instance, collecting indices before a mesh relaxation operation in order to retrieve a specific mesh face after the geometry has dynamically balanced itself according to its internal forces.

2.3.4   On-site Fabrication Chain

2.3.4.1   Fabrication-Parallel-Construction

In recent years, the concept of design-to-production has blurred the boundaries between design and fabrication, but still, the construction phase remains not fully connected.

The bridge between CAM and CAD systems has facilitated the execution and programing of fabrication by the designers, and allowed digitally controlled manufacturing to reach new expressions of form making through mass customization.

On building scale, the communication between fabricators and construction workers has never been necessary, normally architects have been the link between them, but fabricators should be part of the assembly phase too, and this could happen if the machines will move directly to operate on-site.

The prototypical architectures in this research focus on investigating on-site collaborative processes by gathering the various operations in the same location, and because of the tide span of time available, the fabrication, material treatment and assembly processes were designed to run in parallel in order to further optimize the workflow.

The adopted seamless chain design–production–construction integrates a new shift, which can technically be called fabrication-parallel-construction, where the fabrication happens on-site along with the assembly. When a piece goes missing or is damaged, it was replaced in a matter of hours. The advantages are numerous, transport arrangements of the manufactured elements are cut to zero, and labelling, grouping and sorting are organized according to the assembly sequence. All the tasks are directly made on-site, and teams from design, fabrication and assembly are linked not only virtually but also physically collaborating and solving doubts instantly.

The Woven Wood structure employed 104 different plywood pieces to cover a surface of 22.5 m^2, where additional plates and bolts (two types of bolts with a total of 650) were introduced to linearly connect the strips edges. The plywood parts were manufactured from fifteen 1.43×2.44 m boards, and the design, fabrication and assembly processes were completed in just a week (Figs. 2.50 and 2.51).

The Synthesis of Strip Pattern prototype was assembled out of 248 geometrically unique plywood strips and connecting plates fastened together with 1066 bolts,

Fig. 2.50 Woven Wood, sub-strips assembly process

Fig. 2.51 Woven Wood, strips assembly process

allowing disassembly and relocation of the artefact (Fig. 2.52). The surface area was 35 m^2 and was designed and built in the span of 1 week using thirty 1.5 × 3 m boards.

The Laminate Pavilion counts on 302 plywood unique lamellas cut out of 65 boards with the size of 2.44 × 1.22 m, connected permanently with industrial staplers of ½ and 1 inch. The entire fabrication and construction processes of the structure, made out of two 45 m^2 plywood layers, run in parallel over the span of just a month (Figs. 2.53 and 2.54).

The design process for these projects has a series of parameters in common such as strip width, shifted segmentation of the surface and joint size, which delineate the design language for the user. Digitally, material and machine constraints (as wood grain direction, material size available on market and CNC machining space) are embedded in the computational model and integrated as active design drivers to define the strip dimensions. At the micro-scale, the joint design has been defining not only the assembly typology but also the CNC fabrication process.

The plywood strips were CNC manufactured as planar elements. In the first two case studies, they were done so independently from the geometrical complexity with only simple two-dimensional cut (Fig. 2.55), while with the Laminate Pavilion, the cutting process employed a flip mill three-axis CNC. The more complex the fabrication strategy becomes, the more the manufacturing time of the elements increases. Machining processes do not necessarily influence the precision of the assembled structure; the average tolerance in all the three projects varies between ½ mm and 4 mm.

Fig. 2.52 Synthesis of Strip Pattern, strip fastening

Fig. 2.53 The Laminate Pavilion, assembly of the first layer

In order to verify the parallel workflow, a mock-up of each project was realized, testing the process from the fabrication to the assembly. Estimated cutting time,

Fig. 2.54   The Laminate Pavilion, staple gun assembly of 9 mm plywood strips

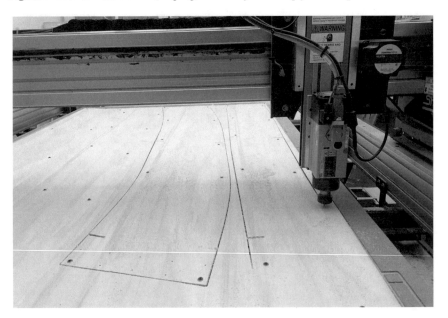

Fig. 2.55   Woven Wood, CNC cutting of 3 mm plywood sheets

material treatment and assembly were calculated for efficiency and effectiveness in the short time span available. Thanks to the on-site fabrication-parallel-construction

Fig. 2.56 Synthesis of Strip Pattern, lightweight and robust structure

approach, despite eventual machining and assembly delays, it was possible to shift and overlap the course of the tasks in order to fit within the estimated deadline constraints.

2.3.4.2 Lightweight Timber Structures

Plywood lamellae are greatly suitable for lightweight architecture; given their elastic bending properties they can be applied through active bending to perform as tough thin plate structure. Lamellar formations can be designed through an intelligent use of the material to gain stiffness and yet be lightweight (Fig. 2.56). Engineering strategies to form-find geometries are key in minimizing the use of material and to perform remarkably on load-bearing cases.

Structural simulations are integrated to preview the deformation, displacement and buckling, by not avoiding but addressing these aspects in favour of the design (Fig. 2.57).

Lightweight structures are commonly simple and quick to assemble, which makes the construction process convenient to execute directly on site. Prefabrication is also a good option as the transport is generally quite flexible thanks to their minimal weight.

The research carried out over these years combined active bending with weaving, sliding components and lamination methods. These techniques focused on smooth curvature transition through strip subdivision and exploring direction of forces in

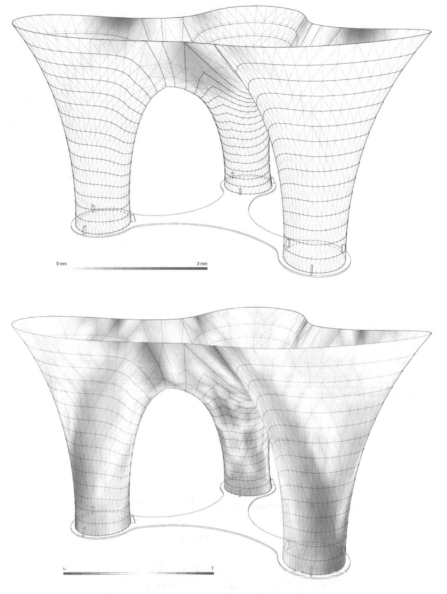

Fig. 2.57 Synthesis of Strip Pattern, displacement and utilization diagrams

order to describe complex, curvilinear, self-supported surfaces without any sub-structure.

Computer-aided architectural design was exploited in favour of digital morphologies, to structurally optimize local and global geometrical force behaviour and to offer a much wider range of design opportunities and geometrical iterations.

Fig. 2.58 Woven Wood, materiality and expression of local twisting

In the three study cases, "the global shape provides additional stiffness, as the pronounced double curvature helps avoiding undesirable deformation modes to the structure".[32]

The Woven Wood aimed to design a wall with differentiated thickness through interlaced strip elements, using global displacement analysis to inform the wall section in order to arrange a gradient of stiffness and flexibility. The strip torsion achieved through shifting the joint pattern, provided additional levels of rigidity and geometrical differentiation, it increased the stiffness locally but also restricted the design possibilities to the limits of the unrollable geometries (Fig. 2.58).

The Synthesis of Strip Pattern began from the meso-scale, with a self-stable component system, based on the ring-like typology and the active bending principle. The resulted structure acquired local self-supporting capacity increased by the double curvature within the global shape. The architectural context addressed merging the design of columns and ceiling together in a smooth continuous surface to demonstrate how conventional, spatial elements can be blended computationally towards ultralight and tough architecture. In the generation of the hyperbolic geometry of the entire structure, a globally continuous strip pattern emerges, defined by the equilibrium state of the oriented forces.

In the Laminate Pavilion, the system takes lamination as strengthening principle where just the first layer is necessary to assemble the essential structure, while the

---

[32]La Magna, R., Bend9 Bending-Active design at Pier 9. 2016. Last accessed on 24 January 2018. http://www.itke.uni-stuttgart.de/entwicklung.php?lang=en&id=76.

Fig. 2.59 Synthesis of Strip Pattern, differentiated joint arrangement

additional layers reinforce it on both strength and global stiffness. The lamination process employs the possibility to design additional skin levels by setting layer differentiation by layer direction and reduce the deformation under secondary vertical loads. In Woven Wood and Synthesis of Strip Pattern, the structural analysis was applied through the simulation of the structure's self-weight alone, while for the Laminate Pavilion it was calculated to support additional live loads, such as to resist the weight of a person.

In this last project, the stabilizing principle is given by the two strip ends, which are anchored to the shell base and the edge, which tangentially connects the strip to the previous one.

Commonly, the joints represent the local weakness of the structure, therefore in these projects, the strip joints are arranged in a shifted manner and distribute these areas, reducing the possibility of failure and showcasing new aesthetic qualities (Fig. 2.59).

The structural analysis helped optimize the stability from the local level in the Woven Wood project. In the Synthesis of Strip Pattern, analysis of displacement was so low that the realized prototype was deformed by less than a centimetre. In the Laminate Pavilion, there was an expected displacement at the shell entrances of about 7 cm, the self-weight of the shell acted in compression on the upper part of the entrance, lowering it down. These margins were acceptable for the designers, and it was decided to displaying the building system alone instead of going adding frames to align the shell edge.

The Woven Wood wall was built from 3 mm thin plywood and spanned about 8 m with a variable height from 1.5 to 3 m. In the Synthesis of Strip Pattern, the structure was made with 3 mm thickness plywood, which covered around 13.7 m^2 with a total weight of less than 70 kg. The Laminate Pavilion was realized out of about 2 cm thickness timber composite, which covered an area of 30 m^2 and had a weight of only 284 kg.

Structures with such statistics highlight the potential of reproducing the system on real building cases with proven structural performance, cost-effectiveness and construction possibilities.

2.3.5 Conclusions

The presented research-oriented work demonstrates the potential of strip applications in timber construction through computational design strategies, digital fabrication and enhanced material integration.

The employment of active bending introduces improvements on mechanical properties and damage tolerant characteristics realizing extremely robust and lightweight structures. Additionally, structural benefits from composite timber lamination have been proved to be effective.

Timber is a workable and transformable material suitable for CNC machining processes, sustainable and natural. The investigation of material enhancement through hygroscopic properties makes the material even more versatile and applicable in new contexts.

The architectural potential manifests itself in the materialization of double-curved topologies, novel spatial qualities and the aesthetics of the surface morphology.

These construction methods can be considered as alternative and efficient fabrication systems to common falsework methods for shells and buildings in general.

Besides the numerous characteristics named on this research, other advantages include thermal insulation and acoustic properties, both of which are not yet fully integrated in the projects main scope.

The full-scale prototypes prove the effectiveness on small-scale structures, whereas the multi-scalar features of the system can be applied to larger scales, opening up the opportunities to building applications with integrated timber strip performances.

References

Adriaenssens S, Block P, Veenendaal D, Williams C (2014) Shell structures for architecture: form finding and optimization. Routledge, New York

Ahlquist S, Menges A (2015) Materiality and computational design emerging material systems and the role of design computation and digital fabrication. In: Kanaani M, Kopec D (eds) The routledge companion for architecture design and practice established and emerging trends. Routledge, London, pp 149–168

Beorkrem C (2013) Material strategies in digital fabrication. Routledge, New York

D'Acunto P, Kotnik T (2013) AA/ETH-Pavilion. In: Proceedings of the TENSINET symposium 2013, Istanbul, pp 99–108

La Magna R, Schleicher S, Knippers J (2016) Bending Active Plates. In: Advances in architectural geometry 2016, vdf Hochschulverlag AG an der ETH, Zürich, p 174

Menges A (ed) (2012) Material computation higher integration in morphogenetic design. Wiley, Hoboken

Menges A (2015) Material synthesis fusing the physical and the computational. Wiley, London

Menges A, Schwinn T, Krieg OD (eds) (2017) Advancing wood architecture a computational approach. Routledge, New York

Pottmann H, Asperl A, Hofer M, Kilian A, Bentley D (2007) Architectural geometry. Bentley Institute Press

Sheil B (ed) (2012) Manufacturing the bespoke making and prototyping architecture. Wiley, Chichester

Thomas KL (2007) Material matters architecture and material practice. Routledge, London

Willmann J, Gramazio F, Kohler M, Langenberg S (2013) Digital by material. In: Brell-Çokcan S, Braumann J (eds) Robotic fabrication in architecture, art and design 2012. Springer Wien, New York, pp 12–27

Printed in the United States
By Bookmasters